Become a Competent Music Producer in 365 Days

Become a Competent Music Producer in 365 Days is a comprehensive, step-by-step guide to the fundamentals of music-production. Over the course of a year, this book takes the reader through ten chapters covering mixing, equalization, compression, reverb, delay and modulation, automation, vocals, synthesis, and mastering.

To combat the patchy nature of 'fast' online content, this book provides an accessible and easily digestible course. Each chapter is broken down into daily readings and tasks, so that each topic can be fully explored, understood, and implemented before moving onto the next, with a range of online video tutorials that offer useful companion material to the book.

Become a Competent Music Producer in 365 Days is an ideal introduction for beginners of all backgrounds, and students in further and higher education music-production classes, as well as aspiring professionals, hobbyists, and self-taught producers, who wish to have a thorough grasp on all the fundamental topics that any experienced music producer should know.

Sam George – The Producer Tutor is a songwriter, composer, producer, and educator from the UK. He's trained in classical and popular music and is a self-taught music producer and audio engineer. From his cutting-edge recording studio in Spain, he collaborates with artists globally.

"Sam George has broken down the fundamental elements of music production and presented them in an easily digestible format. A must read for any producer at any level."

Dom Brown, musician, songwriter, producer, and lead guitarist (*Duran Duran*)

"I love how this book brings effortless simplicity to the art of music production. I fully recommend it."

Damian Keyes, Educator, Founder of *BIMM* and *DK-MBA*

"A bite-sized, action-based, and investigative approach which allows readers to apply industry-standard knowledge and techniques to their creative products and projects. This book is high on my recommended reading list, and it should be on yours too!"

Nathan Lilley, Music Educator, Innovator, and *ELAM* Vice Principal

Become a Competent Music Producer in 365 Days

Sam George

LONDON AND NEW YORK

Designed cover image: © Ninoon via Getty Images

First published 2023
by Routledge
4 Park Square, Milton Park, Abingdon, Oxon OX14 4RN

and by Routledge
605 Third Avenue, New York, NY 10158

Routledge is an imprint of the Taylor & Francis Group, an informa business

© 2023 Sam George

The right of Sam George to be identified as author of this work has been asserted in accordance with sections 77 and 78 of the Copyright, Designs and Patents Act 1988.

All rights reserved. No part of this book may be reprinted or reproduced or utilised in any form or by any electronic, mechanical, or other means, now known or hereafter invented, including photocopying and recording, or in any information storage or retrieval system, without permission in writing from the publishers.

Trademark notice: Product or corporate names may be trademarks or registered trademarks, and are used only for identification and explanation without intent to infringe.

British Library Cataloguing-in-Publication Data
A catalogue record for this book is available from the British Library

Library of Congress Cataloging-in-Publication Data
Names: George, Sam (Music producer), author.
Title: Become a competent music producer in 365 days / Sam George.
Description: New York : Routledge, 2023. | Includes bibliographical references and index. | Identifiers: LCCN 2022057003 (print) | LCCN 2022057004 (ebook) | ISBN 9781032446110 (paperback) | ISBN 9781032446141 (hardback) | ISBN 9781003373049 (ebook)
Subjects: LCSH: Sound recordings—Production and direction. | Popular music—Production and direction.
Classification: LCC ML3790 .G46 2023 (print) | LCC ML3790 (ebook) | DDC 781.49—dc23/eng/20221202
LC record available at https://lccn.loc.gov/2022057003
LC ebook record available at https://lccn.loc.gov/2022057004

ISBN: 978-1-032-44614-1 (hbk)
ISBN: 978-1-032-44611-0 (pbk)
ISBN: 978-1-003-37304-9 (ebk)

DOI: 10.4324/9781003373049

Typeset in Goudy
by Apex CoVantage, LLC

Access the Support Material: www.theproducertutor.com

Contents

Preface   xvi
Acknowledgements   xix

1   Balancing a mix   1

    Day 1 – Welcome   1
    Day 2 – Balancing a mix   1
    Day 3 – Why is balancing a mix so overlooked?   2
    Day 4 – SPL vs. perceived loudness   3
    Day 5 – 24-bit   3
    Day 6 – Reference tracks   4
    Day 7 – What is gain staging?   5
    Day 8 – Normalising region gain   7
    Day 9 – Pre-fader metering   8
    Day 10 – Phase cancellation   9
    Day 11 – Bottom-up mixing   9
    Day 12 – Top-down mixing   10
    Day 13 – The clockface technique   11
    Day 14 – Mono compatibility   13
    Day 15 – The VU meter trick   13
    Day 16 – Balancing vocals   14
    Day 17 – The mixing hierarchy   15
    Day 18 – A singular focal point   15
    Day 19 – The rule of four   16
    Day 20 – The pink noise technique   16
    Day 21 – Loudest first   17
    Day 22 – Set a timer   18
    Day 23 – Leave your busses   18
    Day 24 – Watch your meters   18
    Day 25 – Split them up   19
    Day 26 – FFT metering   20

Day 27 – Metering checklist 20
Day 28 – Unit summary 21
Checklist 21
Further reading 21

2 Panning a mix   23

Day 1 – Panning a mix 23
Day 2 – Know your pan pots 23
Day 3 – From whose perspective are you mixing? 24
Day 4 – Kick and snare 25
Day 5 – Overheads and rooms 26
Day 6 – Hi-hats 26
Day 7 – Toms 27
Day 8 – Bass 28
Day 9 – Guitars 28
Day 10 – Synths and keys 29
Day 11 – Horns, wind, and strings 30
Day 12 – Vocals 31
Day 13 – The LCR technique 31
Day 14 – Low-frequency panning 32
Day 15 – Double-tracked guitars 33
Day 16 – Complementary panning 33
Day 17 – Snare on- or off-centre? 34
Day 18 – Narrow verses, wide choruses 35
Day 19 – Automating movement 35
Day 20 – Checking mono compatibility 36
Day 21 – Consider the club 37
Day 22 – Crosstalk 37
Day 23 – Don't overload one side 38
Day 24 – Recreating a vintage vibe 38
Day 25 – Less is more 39
Day 26 – Orchestral panning 1 39
Day 27 – Orchestral panning 2 40
Day 28 – Unit summary 41
Checklist 41
Further reading 42

3 EQ   43

Day 1 – Applying EQ 43
Day 2 – Fundamentals and overtones 43

Day 3 – Defining frequency bands 44
Day 4 – Different types of equalisers 45
Day 5 – Anatomy of an EQ 1 47
Day 6 – Anatomy of an EQ 2 48
Day 7 – Subtractive/corrective EQ 49
Day 8 – Additive/creative EQ 49
Day 9 – EQ can't fix a poor recording 50
Day 10 – Have intent 51
Day 11 – Learn the frequency spectrum 51
Day 12 – Use an EQ chart 52
Day 13 – The big four: Remove the gross stuff 52
Day 14 – The big four: Enhance the good stuff 53
Day 15 – The big four: Make things sound different 54
Day 16 – The big four: Create space 55
Day 17 – Don't be fooled by volume 55
Day 18 – Small for natural, large for unnatural 56
Day 19 – Prioritise cuts 56
Day 20 – Don't EQ in solo 57
Day 21 – Small changes add up 58
Day 22 – Subtle with stock, bold with analogue 58
Day 23 – Don't obsess over plugin order 59
Day 24 – Remove the mud 59
Day 25 – EQ in mono 60
Day 26 – Dynamic EQ 61
Day 27 – How do you pick your EQ? 61
Day 28 – Unit summary 62
Checklist 63
Further reading 63

4 Compression 64

Day 1 – What is compression? 64
Day 2 – Compressor anatomy: Ratio 65
Day 3 – Compressor anatomy: Threshold 65
Day 4 – Compressor anatomy: Attack and release 66
Day 5 – Compressor anatomy: Knee and make-up gain 67
Day 6 – The four analogue architectures: VCA 68
Day 7 – The four analogue architectures: Optical 68
Day 8 – The four analogue architectures: FET 69
Day 9 – The four analogue architectures: Variable Mu 69
Day 10 – Use one compressor to learn 70
Day 11 – What comes first: Compression or EQ? 71

Day 12 – Don't compress just because you can 71
Day 13 – Parallel compression is the best of both 72
Day 14 – Matching input and output levels accurately 73
Day 15 – Avoid presets and solo 73
Day 16 – Stacking compressors 74
Day 17 – Compressing low end 75
Day 18 – The extreme-threshold technique 75
Day 19 – Don't kill the transients 76
Day 20 – A little here, a little there 76
Day 21 – The right compressor for the job 77
Day 22 – Don't be fooled by volume 78
Day 23 – Complementing compression with transient enhancement 78
Day 24 – Sidechain compression 79
Day 25 – Sidechain EQ 80
Day 26 – Colouring your sound 81
Day 27 – Multiband compression 82
Day 28 – Unit summary 82
Checklist 83
Further reading 83

5   **Reverb**   85

Day 1 – What is reverb? 85
Day 2 – Why do you need reverb? 85
Day 3 – What does reverb do? 87
Day 4 – Different types of reverb: Hall 88
Day 5 – Different types of reverb: Chamber 88
Day 6 – Different types of reverb: Room 89
Day 7 – Different types of reverb: Plate 89
Day 8 – Different types of reverb: Spring 90
Day 9 – Different types of reverb: Convolution 91
Day 10 – Different types of reverb: Gated 91
Day 11 – Different types of reverb: Honourable mentions 92
Day 12 – Reverb parameters: Type, size, decay, and mix 93
Day 13 – Reverb parameters: Pre-delay, early reflections, and diffusion 94
Day 14 – Eight steps to perfect reverb: Aux send or insert slot? 95
Day 15 – Eight steps to perfect reverb: Selecting reverb type 95

Day 16 – Eight steps to perfect reverb: Set your size 96
Day 17 – Eight steps to perfect reverb: Set your decay 96
Day 18 – Eight steps to perfect reverb: Set your pre-delay 97
Day 19 – Eight steps to perfect reverb: Set your early reflections 98
Day 20 – Eight steps to perfect reverb: Set your diffusion level 98
Day 21 – Eight steps to perfect reverb: Adjust the dry/wet/send amount 99
Day 22 – Top tips: Less is more 99
Day 23 – Top tips: Mono, stereo, or panned? 100
Day 24 – Top tips: Pre- or post-fader? 101
Day 25 – Top tips: Treat reverb like an instrument 101
Day 26 – Top tips: How many reverbs should you use? 102
Day 27 – The formula 103
Day 28 – Unit summary 104
Checklist 105
Further reading 105

6 **Delay and modulation effects** 107

Day 1 – What is delay in music-production? 107
Day 2 – Delay times 107
Day 3 – Primary delay types: Tape 108
Day 4 – Primary delay types: Analogue 109
Day 5 – Primary delay types: Digital 109
Day 6 – Delay subtypes: Slapback echo 110
Day 7 – Delay subtypes: Doubling echo 111
Day 8 – Delay subtypes: Looping 111
Day 9 – Delay subtypes: Multitap 112
Day 10 – Delay subtypes: Ping-pong 112
Day 11 – Delay subtypes: Dub 113
Day 12 – A reminder about phase 114
Day 13 – Modulated delay: Chorus 114
Day 14 – Modulated delay: Flanger 115
Day 15 – Modulated delay: Phaser 116
Day 16 – Top tips: Have an intention 116
Day 17 – Top tips: Be selective 117
Day 18 – Top tips: Work out your timing 118
Day 19 – Top tips: Leakage 119
Day 20 – Top tips: Processing delays 120

Day 21 – Top tips: Automation 121
Day 22 – Top tips: Create manual delays 121
Day 23 – Top tips: Flamming delays 122
Day 24 – Top tips: Panning delays 122
Day 25 – Top tips: Tempo-syncing 123
Day 26 – Top tips: Making sounds bigger/wider 124
Day 27 – Top tips: Enhance important things 124
Day 28 – Unit summary 125
Checklist 126

7 Automation 127

Day 1 – What is automation? 127
Day 2 – Automation modes 127
Day 3 – Automation types: Fades and curves 128
Day 4 – Automation types: Binary, step, and spike 129
Day 5 – Have intent! 130
Day 6 – Bouncing automation 130
Day 7 – Three stages: Problem-solve 131
Day 8 – Three stages: Flow 132
Day 9 – Three stages: Create 132
Day 10 – When should you automate? 133
Day 11 – Use No. 1: Levels 134
Day 12 – Use No. 2: Stereo width 135
Day 13 – Use No. 3: More automation, less compression 136
Day 14 – Use No. 4: Fix plosives and sibilance 137
Day 15 – Use No. 5: Increase plugin control 138
Day 16 – Use No. 6: Aux sends 138
Day 17 – Animate your transitions 140
Day 18 – Automate synth parameter changes 140
Day 19 – Emphasise drum fills 141
Day 20 – Tidy up 142
Day 21 – Accentuate vocals 143
Day 22 – Automate panning for sound design 144
Day 23 – Automate EQ 144
Day 24 – Make e-drums sound more natural 145
Day 25 – Tempo-synced automation effects 146
Day 26 – Automate your master buss 147
Day 27 – Automate your tempo 148
Day 28 – Unit summary 149
Checklist 150

8 Vocals — 152

- Day 1 – Vocals 101 152
- Day 2 – Create space pockets 153
- Day 3 – Open-back vs closed-back 153
- Day 4 – Use a pop shield 154
- Day 5 – Consider mike emulation 155
- Day 6 – DSP-powered monitor effects 156
- Day 7 – The proximity effect 156
- Day 8 – Get your levels right 157
- Day 9 – Get your mix right 158
- Day 10 – Record everything! 158
- Day 11 – Harmonies and layers 159
- Day 12 – Gate post-recording 160
- Day 13 – Loop and comp 161
- Day 14 – Tuning and other modulation effects 162
- Day 15 – Subtractive EQ 163
- Day 16 – Consider the genre 164
- Day 17 – Tone-shaping EQ 164
- Day 18 – Compression 165
- Day 19 – More compression 166
- Day 20 – . . . and even more compression! 167
- Day 21 – De-ess with care 167
- Day 22 – Saturation and distortion 168
- Day 23 – Don't overdo reverb and delay 169
- Day 24 – Pan backing vocals 169
- Day 25 – Converting to MIDI 170
- Day 26 – Vocal automation 171
- Day 27 – Frequency allocation 172
- Day 28 – Unit summary 173
- Checklist 173
- Further reading 174

9 Synthesis — 175

- Day 1 – Synthesis terrified me 175
- Day 2 – What is a synthesiser? 176
- Day 3 – Oscillators 176
- Day 4 – Waveshapes 178
- Day 5 – Combining oscillators 179
- Day 6 – Tuning, unison, and voices 180
- Day 7 – Filters 181

Day 8 – Amplifiers and envelopes 181
Day 9 – Modulation 183
Day 10 – LFOs 184
Day 11 – Other possible features 184
Day 12 – What makes a synth unique? 186
Day 13 – 1/10: Subtractive synthesis 186
Day 14 – 2/10: FM synthesis 187
Day 15 – 3/10: Sample-based synthesis 187
Day 16 – 4/10: Wavetable synthesis 188
Day 17 – 5/10: Vector synthesis 188
Day 18 – 6/10: Additive synthesis 189
Day 19 – 7/10: Spectral synthesis 189
Day 20 – 8/10: Physical modelling 189
Day 21 – 9/10: Granular synthesis 190
Day 22 – 10/10: West Coast synthesis 190
Day 23 – The main synth sounds 191
Day 24 – 20 top tips: 1–5 192
Day 25 – 20 top tips: 6–10 193
Day 26 – 20 top tips: 11–15 194
Day 27 – 20 top tips: 16–20 195
Day 28 – Unit summary 196
Further reading 196

10 Mastering 197

Day 1 – The dark art 197
Day 2 – Mixing vs mastering 197
Day 3 – A brief history 198
Day 4 – Preparing to master 199
Day 5 – Mix levels and compression 200
Day 6 – Loudness, the war, and penalties 201
Day 7 – The most important piece of gear is . . . 202
Day 8 – Don't master your own tracks 203
Day 9 – Digital vs analogue 204
Day 10 – Optimising, not fixing 205
Day 11 – Good mastering preparation 206
Day 12 – The four stages of mastering: EQ 206
Day 13 – The four stages of mastering: Compression 208
Day 14 – The four stages of mastering: Limiting 209
Day 15 – The four stages of mastering: Metering 210
Day 16 – Using appropriate reference tracks 211

Day 17 – Trimming the start and end 212
Day 18 – Bouncing/rendering/exporting 213
Day 19 – Metadata 214
Day 20 – Beyond the essentials: Saturation 215
Day 21 – Beyond the essentials: Stereo enhancement 216
Day 22 – Beyond the essentials: Parallel processing 217
Day 23 – Beyond the essentials: Serial limiting 218
Day 24 – Adding reverb 218
Day 25 – Avoid ear fatigue 219
Day 26 – Stem mastering 220
Day 27 – Considering the medium and technical limitations 221
Day 28 – Unit summary 222
Further reading 223

Index 224

Preface

'Music producer' is a term that has evolved over time. In the traditional sense, music producers assist artists to make records. They help bring artists' visions to life. This can involve technically and creatively guiding them, and may include anything from coaching the artists through their performances to organising meetings, scheduling, and budgeting. However, it tends to mean something a little different when referred to in the modern day. When people call themselves music producers nowadays, they simply mean people who create music. Literally, they produce music. This may involve artist liaison, organising sessions, and all those other things, but often the producers may themselves be the artists. This is the context within which I operate throughout this book. I'm referring to the music producer, a self-sufficient entity that creates music for themself or for others.

I have spent a lot of time watching online educational content over the years. As a music producer, I am entirely self-taught. My journey began when I started writing songs when I was 13. I started in the traditional sense with pen and paper, but by the time I was 16, I was making sketchy demos using Cubase. By the time I got to university, my demos were acceptable at best. My degree was based primarily on songwriting and contained little production time. But this is where my interest in the subject really sparked. I spent hours upon hours making demos and sending them back and forth between my co-band members. Over time they improved but were nowhere near professional in quality. Bear in mind that I started university in 2005. This is roughly the time Facebook launched. At that time, you needed to be affiliated with a university (or college) to join. YouTube wasn't on the scene at all. As online content exploded over the following ten years, I began consuming as much of it as possible.

Fast forward to now, and I've watched just about every so-called educational content creator there is. Out of them all, perhaps only a handful know what they're talking about. The vast majority are either factually incorrect, only partially correct, or take so long to give you the information you need that you'll have moved on to something else by the time they get there.

Having become sick of nonsense content online, I decided to launch The Producer Tutor. I've spent months, if not years, looking for the correct information, learning it, and translating it into an understandable format.

And that's what this book is about. It's about giving you the information you need concisely, precisely, and accessibly in a way that you can easily refer to. I've compiled the content I believe will make you a competent music producer and have left out the unnecessary, overcomplicated waffle. What you'll find in this book is cover-to-cover value.

So, who's this book for? It's for beginners. Those who have only just started dabbling in the world of music-production. It's for intermediates. Those that have been at it for a while. And it's for more advanced, more experienced users too. A basic working knowledge of a DAW of your choice is a prerequisite, but that's it.

Most importantly, it's for those that want to learn music-production properly. If you want to understand the subject thoroughly, not superficially, and if you're going to be well informed and not patchy in your knowledge, then this book is for you. There are no shortcuts, no quick hacks. The information in these pages is thorough, logical, and structured but delivered in a condensed way that should be understandable. Note that I'm not going to tell you what frequencies to cut or boost from your kick drum or how to compress your vocal. Those questions are unanswerable. But I will introduce you to tools and techniques that will allow you to explore those topics in an informed manner.

I've purposely ensured that each 'day' is short. I've labelled the sections of chapters as days because I want you to read one a day and not binge through a whole load. I want you to read a day, go to your DAW and apply the information you've just learned. Each day has a task to help you focus on the relevant subject matter. From my experience as an educator, students learn best when they have time to apply their new knowledge. By delivering the information you want in bite-sized chunks, I'm ensuring you can reread sections as many times as you need to ensure you've understood the content. Given this format, you may be expecting 365 individual days of content. This isn't the case. There are ten chapters of 28 days each, leaving about two months over at the end. Once you've completed a chapter, I assume you'll want to spend a few days applying everything from that section. With all those additional days added together, we get to roughly 365. I've also left you a list of things to check off after each unit to ensure you're exploring everything in detail.

It's important to clarify from the outset that, whilst I will cover many mixing and mastering aspects, this book is not in itself a guide to mixing or mastering. This book is a guide to producing music, the tools available to you and how you can use them effectively.

With all that said, let's dive in!

Further digital resources can be found here:

- www.theproducertutor.com
- www.youtube.com/@theproducertutor.com
- www.tiktok.com/@theproducertutor
- www.instagram.com/theproducertutor

Acknowledgements

My thanks must begin with Mum and Dad. Their unwavering support truly knows no limit. They have guided and supported me practically and financially through every challenge, of which I have posed them many. With lesser parents this book wouldn't exist. Perhaps I wouldn't either. Mum and Dad, words can't describe my gratitude for everything.

My wife, Estrella. You have shown complete faith in me, even when my own wavered. You've supported me in my risk-taking and have picked me up every time I've fallen. I owe you everything.

My secondary school music teacher, David Leveridge. You tried your best to understand me and support my creativity, even when it went against the grain. I was a challenging student, sometimes for the right reasons, sometimes not. But you backed me. Good education begins and ends with the teacher, and you were one of the best.

My brothers in crime, Tim Talbot, Jay Armstrong, John Atkins, and Harry Armstrong, the Armstrong boys. The days, months, and years on the road with you in sticky-floored clubs and rural recording studios moulded my popular music education. I learned how to write good songs with you, learned how to record music with you, learned how to tour with you. You'll always be my first true love.

The two best role models any teacher could ask for – Liz Penney and Alice Gamm, The BRIT Queens. Liz, you gave me my shot, took a punt on me when you already had a safe option. You let me get my foot in the door at BRIT. Alice, you showed faith in me, trusted me, gave me responsibility, and let me run with it. The two of you are truly wonderful educators. I learned so much from you both in such a short time. Thank you.

Dec Cunningham and Mat Martin. It's hard to quantify how much knowledge I've robbed from you both. Daily I'd be in one of your ears for something or another. Never has knowledge been given up so willingly. So much of what I know now I gleaned from you two. You're both legends.

If the Armstrong boys were my first true love, Nathan Lilley, you were my second. A wonderful friend, a sickeningly talented musician, the most passionate educator with a fierce commitment to the students, and a fountain of knowledge in all things music. Thank you for your friendship, and for answering my WhatsApps.

And finally, my oldest friend, Luke Fox. We message daily, almost always about music. Your counsel and interest in exploring the subject from every angle continually inspires me to review and reassess. Thank you for challenging me.

Unit 1

Balancing a mix

Day 1 – Welcome

Are you tired of trawling through YouTube looking for thorough, detailed information on a music-production topic? Are you even sure that the information you're finding is of good quality?

I'm Sam, the Producer Tutor – your musical PT. I'm here to bring you my 365-day course that'll take you from enthusiastic amateur to competent producer.

When I was learning music-production, the available content was sparse and generally not that engaging. Nowadays, the issue is compounded by every other home producer filling the internet with tutorials on creating different sounds and effects. But nobody's talking about the fundamentals. The basics. The foundations of music-production.

That's what I promise to give you. A structured, thorough approach to learning music-production properly. I'll teach you from the ground up, covering all the fundamental topics you'll need to digest to solidify your production skills.

What makes me qualified to teach you? I spent six years teaching songwriting and music-production at The BRIT School in South London. Many of my students have achieved success nationally and internationally. Some have even gone on to sell platinum records.

So, strap yourself in, and have your notepad at the ready because, within 365 days, I'll have taught you just about everything you need to know to produce music competently.

Day 2 – Balancing a mix

So here we go: Unit 1. We're going to start this process by looking at mixing. And for the whole of the first Unit, I'm going to be teaching you all about balancing a mix. That's right, an entire month focusing on balancing a mix. That's how thorough we're going to be in this course! You might wonder how there's possibly enough within this topic to warrant spending a whole month on it. But I promise you that you will feel so much better equipped and informed when you get to the end of this unit than you do now.

For me, there are four fundamental skills that are the building blocks of producing music. These are balancing a mix, panning a mix, applying EQ, and applying compression. If you can master these four elements, you're 90% of the way towards a great-sounding track. Everyone talks about EQ and compression, sometimes panning, but mix balancing gets very little airtime. And it's *the* most significant part.

> TASK – Find a resource online where you can download free multitrack stems. You'll be given a few different options if you Google 'Free Multitrack Stems'. You'll want various options to use as practice in the coming units.

Day 3 – Why is balancing a mix so overlooked?

Why is it the most overlooked aspect of music-production, then? In my opinion, it's because, technically, it's the most straightforward. All you need to do is make everything in your project sound balanced. How complicated can that be? But this is the stage where you'll make or break your entire mix. If you don't get this right, you'll end up fighting yourself all the way down the line. Get it right, though, and you'll make the rest of your session 100 times easier.

But where do you start? Well, and this is key, you need to start from a clean mix. Starting from a clean mix means you've committed to your instrumental sounds. You've bounced all your MIDI to audio and removed any plugins or processes you were using in your writing session. Whilst I mention it, this is an important distinction to make: Keep your processes separate. You shouldn't attempt to mix your track until you've committed to your song. You're not ready to mix if you're still making compositional decisions.

So, having said this, let's consider why we start with mix balancing. It's simple. Balancing your mix first allows you to reference back to your balanced mix at every stage throughout the process and will allow you to guarantee that the processing decisions you make down the line make your mix better, not just louder. Modern productions often have a lot going on. So, to hear everything clearly, you must start from a balanced starting point.

> TASK – Make a bullet point list of how you would typically begin balancing a mix. I want you to compare with my guidance at the end of this unit to see where you can improve your current flow.

Day 4 – SPL vs. perceived loudness

Let's get technical for a moment. Loudness impacts massively on how we perceive sound. When you play a track at a low volume, the human ear can still easily make out frequencies from 1 – 5kHz, but hearing sound under 500Hz is almost impossible. For this reason, you'll struggle to hear the bass at low levels, but you'll still comfortably hear the vocal. As volume increases, our hearing becomes more balanced. The optimum loudness for your home studio is between 73 – 76dB SPL. But your mixer in your DAW won't have any control over how loud you listen to your music, only how loud each track is within your project. You can use a dB reader on your phone to check the loudness at which you're listening. You can grab a free one from your app store. Make sure you measure the level in your listening position, i.e., where you sit to mix.

> TASK – Use a loudness meter on your phone to calibrate an appropriate SPL level in your listening position.

Day 5 – 24-bit

But before you even reach for a fader, there are two vital things you need to do first. The first thing is to make sure you're working in 24-bit. When digital audio began to take over from analogue, loads of old practices got carried over. One of the biggest was the idea of recording as loudly and cleanly as possible. Why? To keep the signal above the noise floor. What is the noise floor? Well, every electronic device produces noise. These devices include your mikes, cables, and audio interfaces. The noise floor refers to the amount of noise your equipment makes. The noise floor was undoubtedly an issue when working on tape. Using a lot of tracks in 16-bit audio can begin to gather a noticeable noise floor. But 24-bit solves that. The 24-bit noise floor is so low that you can give yourself loads of room (15 – 20dB) between your peaks and 0dBFS without worrying about noise or loss of resolution.

There's another benefit to working in 24-bit. Imagine a ruler, where the centimetre markings are the 16-bit measures, and the 24-bit markings are the millimetres. When using 16-bit you can only measure to the nearest centimetre marking, so your accuracy when measuring could be some way off. When using 24-bit your level of accuracy is much higher. The higher bit depth allows your computer to reproduce subtle variations in the waveform more accurately. There's a lot of complex maths that I'll spare you now, but the important headline here is this: 24-bit digital audio has a maximum dynamic range of 144dB. 16-bit doesn't get close to this, offering just 96dB as its upper limit.

Whatever your DAW of choice is, ensure that you are set up to work in 24-bit.[1]

TASK – Ensure that your DAW is set up to work at a bit depth of 24-bit.

Day 6 – Reference tracks

The second thing you need to do before reaching for a fader is to select some reference material. Producers suffer from a well-known problem. It's called ear fatigue. It means the longer you work on something, the more you listen to something, the more familiar you get with hearing it, the less detail you hear. This deterioration occurs faster if you work at higher volumes. It's an unavoidable thing. Weirder still is that a mix will sound different to you from day to day or session to session based on several factors that affect your ears. The room and environment you work in also enormously alter how your mix sounds. So, what's the best way to negate all these factors? Use reference tracks! Comparing your mix to a professionally produced record will allow you to make informed mix decisions rather than making impulse mix choices without context. Working in this way will also help your mixes translate better between different listening environments. I thoroughly recommend the plugin Reference from Mastering the Mix. It is a super powerful reference plugin that automatically level matches the reference tracks with your project. It is vital to make mix decisions based upon tracks that are the same loudness. Remember what we said earlier about things being louder sounding more balanced and therefore better?

If you have Reference, place it on your stereo out. You can simply drag and drop your reference tracks into the plugin. Ensure that the level match feature is engaged. This feature will ensure that your project's level and reference track's level remain the same, allowing you to make informed mix decisions.

If you don't have Reference, don't fret. You can achieve the same result, with just a little more work. In this case, route everything in your mix to a new buss. You can use this as your mix buss instead of your stereo out. Drag your reference tracks into your project on new tracks but route these to your stereo out, not your new mix buss. Then, place a metering plugin on your stereo out. Use the metering plugin to match the levels of your project and your reference tracks. Focus on matching the loudness of the tracks using your eyes *and* your ears. Use the loudest sections of the songs for this. The peak levels will be quite different as your reference tracks will already have lots of compression and limiting on them.

Now is an excellent time to identify a helpful rule I like to follow. Good plugins fall into one of two categories. Either they make an aspect of producing markedly quicker or they make it easier by reducing your workload. That applies to Reference. Or they allow you to generate sounds that you can't recreate in any other way. Most plugins fall into this category. Plugin manufacturers love to convince you that their latest product is the only one that will allow you to create a specific sound. This is very rarely the case. Now, I'm not against third-party plugins at all. I own a lot of them. But before spending any of your hard-earned cash, do your research. Read some blogs. Watch some YouTube videos. Ensure what you're looking to buy is a good spend and that you're not just being won over by an effective marketing campaign.

Having said all that, you're now ready to start reaching for faders, which is precisely what you'll be doing tomorrow.

TASK – Decide how you're going to reference. If you're going to use a third-party plugin, ensure that you are familiar with how it works. Read the manual! If you're not going to use a designated referencing plugin, practise importing reference tracks into your project and level-matching them with your project.

Day 7 – What is gain staging?

OK, so you're ready to jump into your mix and start balancing everything with your faders, right? Wrong! Before you do anything, you need to gain stage.

Getting a clear and punchy mix isn't just about processing and effects. It begins by setting the correct levels at which you track and mix. If you want to avoid a weak, lifeless, brittle, or harsh mix, gain staging is an integral part of the process.

Now, gain staging is a straightforward concept that many people don't understand or just wholly overlook. The principle is this: Gain staging is the process of ensuring that the level on each of your tracks is healthy, not too hot, and not too quiet. There are several benefits to doing this.

You make your working signal on each channel more consistent by gain staging things. Think about it. The level at which you record each part – your vocals, guitars, drums, etc. – will likely all have been trimmed in at slightly different levels. Especially if you've been working with virtual instruments, by default these will probably have been quite hot in the instrument. Or, if you've tracked parts over multiple recording sessions or days, your trim levels will likely not be precisely the same. By gain staging, you can level the playing field across all your tracks. This process has the knock-on effect of ensuring that you don't need to start making significant moves with your faders to compensate for disparities between channels.

However, the most significant benefit comes when sending your signal through plugins. If your level is too low, you won't take advantage of the digital 24-bit system's full length. Your DAW will fill the unused headroom with nothingness, and when your digital signal is returned to analogue as it comes out of your speakers, it will sound unclear and limp. That sucks. But in fact, it's the opposite end of the spectrum that sounds worse. When you work too hot, it barks and distorts in nasty ways, which tend to attack you in short bursts. You avoid both the limp, lifeless mix and the crunchy, distorted one by gain staging.

So, what's the correct level to gain stage to? Generally speaking, the optimum level at which to send a signal through a plugin is -18dB. Manufacturers design and build plugins to work best with this signal level coming through them. But there are multiple ways of measuring your signal, so what's the best way? Well, your channel meters are what's known as full-scale peak meters. These work super fast and display the highest peak level of the signal on each track. But actually, these meters work much faster than the human ear does, so in terms of gain staging, they aren't the most helpful representation.

What's much more helpful is a VU meter. These have been around since the '30s and work much slower, much more like how the human ear works. Their response time is typically around 300ms. You can calibrate your VU meter to target your chosen dB level. So, if you calibrate your VU meter to target -18dBFS, this will mean that when you're hitting 0dBVU on the meter, you have 18dB of headroom. This method is the best way to gain stage to optimise your signal for processing. This isn't an exact science, so you'll need to proceed with some caution.

Now, you still need to keep your peak meter in mind. You don't want to exceed -10 to -6dBFS at the most, so you must watch that too. But the VU meter method is the best way to gain stage most effectively, and importantly, humanly.

Bear in mind also that sounds with a wide dynamic range will be trickier to gain stage using the VU meter than more dynamically consistent sounds. A simple guideline is to gain stage your percussive elements using the channel full-scale peak meters, trimming up to your chosen peak level (between -10 to -6dB). For anything else, use your VU meter, aiming to hit 0dBVU, and then check your peak meter to ensure you're in the safe zone.

> TASK – Ensure you have a VU meter available within your DAW. If your DAW doesn't have a stock one, install a third-party option.

Day 8 – Normalising region gain

The natural way to want to gain stage is by reaching for the fader on your channel strip and pulling it up or down to get your desired level. But I advise against this. As a general rule, I recommend leaving your faders at unity gain as long as possible. When you come to automate levels much further down the line (a topic that we'll cover later in the course), your faders will all be at sensible levels and not all over the place.

Instead, implement a gain plugin onto the track and use this to trim. Have your VU meter set up on your mix buss. Make sure you've set your reference level in the meter to -18dB and make sure it's operating as a VU meter (some VU plugins will allow you to monitor a range of different things: RMS, PPM, EBUR128). Then go through your tracks one at a time to check your levels. Keep an eye on your peak meters too. As mentioned yesterday, on content with a wide dynamic range they'll peak much higher when you push them on the VU meter. Just check that your VU level is in the ballpark of 0dBVU. You don't need to be too particular about this. You just want to be somewhere close without going over.

Alternatively, you can trim the level of your regions instead. You can do this by individual region or as a whole track at a time. Every DAW will have its own version of this. In Logic Pro you can do it automatically with the Normalize Region Gain function. In Pro Tools you have the Clip Gain function, and so on. You want to target around -10dBFS, meaning that the highest peak on any individual channel will be -10dB. I tend to find this gets you in the ballpark for hitting around 0dBVU but will still leave you some headroom. Setting the peak level of all channels to the same level doesn't mean they'll all be at the same loudness. A track with much higher frequency content that is very transient-heavy will appear much quieter on a VU meter than something low with sustained notes when peak levels are equal. At -10dBFS your hi-hats will hardly tickle the VU meter, whereas your bass will meter strongly. This isn't a problem, but it's worth being aware of otherwise you may find yourself unnecessarily cranking the gain on a hi-hat to get it metering where you want it on your VU meter.

> TASK – Find and practise normalising region gain (or your DAW's equivalent).

Day 9 – Pre-fader metering

The next thing you're going to do is to set up your project to use pre-fader metering. By default, in most DAWs, your project will be set up to use post-fader metering. Many people don't know this, let alone know what that means. But it's simple. Post-fader metering means that the level displayed on the meter of each channel comes after the fader. So, what you do with the fader will affect the metered level. When you're mixing, this isn't very helpful. What you want to do throughout your mixing process is make informed decisions about whether the changes you've made to a sound, be that through EQ, compression, or anything else, is improving the sound and not just making it louder. Remember what we said earlier about louder things fooling us into thinking they sound better? You want to gain stage to match the input and output levels through every plugin you use in your project. This is how you ensure your processing decisions are genuine improvements to your mix.

So, to help you with this, you can set your project to pre-fader metering. This means that the level displayed on the meter comes before the fader, so any changes you make to your fader won't affect the meter level. This is much more helpful for mixing to check that your signal pre-and post-plugins are the same. You may need to do a little bit of digging around in your DAW to find how to turn this on. I know in Logic Pro the option isn't visible by default.

This point of view may be contrary to what you've previously heard. Some will say that pre-fader metering is best for recording, and post-fader is best for mixing, given that the post-fader level is what is being sent to the mix buss. I believe that if you're gain staging correctly and balancing your mix the correct way, you'll never have a problem with the levels you're sending to your mix buss. Therefore, I find it more valuable to have pre-fader levels showing. Ultimately, it's up to you.[2]

> TASK – Make sure you know where and how to toggle between post- and pre-fader metering in your DAW.

Day 10 – Phase cancellation

Let's mix some drums! But wait, you're still not ready to reach for a fader. Before you do, you've got to check for another massively misunderstood phenomenon called phase cancellation. Phase cancellation is an audio phenomenon where the waves of multiple tracks work against each other to eliminate specific frequencies. The resultant sound is often flat and dull. This will usually happen when you've used multiple microphones to record one instrument. It's most common on drums, where you will often use multiple mikes on the drum kit. It generally happens when the sound source's mikes' distance isn't the same, so the source's sound reaches the mikes at slightly different times. When things are out of phase, it means that whilst one sound is at a peak, the other is in a trough. The result is that the signals partially cancel each other out, leaving you with that flat and dull sound I mentioned. The simple fix here is to flip the polarity of one of the channels. This turns the waveform upside down and will therefore be in phase rather than out of phase. This means that the signals will now add to, rather than subtract from, each other.

Because you've placed gain plugins on your channel strips to trim, the great news is that you can use these to check for any phase issues. You simply hit the polarity button, which will flip the phase. You'll hear immediately if you are creating or eliminating phase issues. One will sound full-bodied, and one will sound limp. It's obvious which one you're after!

Check for this on anything that contains multiple mikes on the same sound source. So, think kick drum, snare drum, overheads, guitar amps if you're using more than one mike, anything you've recorded in stereo such as a piano, and so on.[3]

TASK – Practise toggling the polarity on your channel strips. Ensure you are clear on the difference in sound between something that is in phase vs out of phase.

Day 11 – Bottom-up mixing

OK, now let's reach for some faders! You're going to start with what's called a bottom-up approach. Find the fullest section of your project and loop it leaving your snare top or main snare sound at unity. Bring everything else all the way down. You're going to build your drum sound around your snare because, typically, the snare is the foundation of the backbeat and, therefore, is one of the mix's loudest elements (if you're working in a style where the kick is more important than the snare, then bring that up first instead). If you have a snare bottom or other snare sounds, bring these up to taste around your main snare sound.

Then bring your kick tracks up so that they're almost as loud as your snare. The low end should feel full and robust without interfering with the snare drum's bottom end. Next, you'll bring in your toms. If used sparingly, they can be almost as loud as your snare. But if they're used a lot, they should sit a little lower in the mix.

Now bring in any cymbals, overheads, and room mikes you may have as required. How you do this will vary depending on the genre you are producing, so pay attention to your reference tracks to identify a reasonable level. They should support your spot mikes rather than overpowering them, so pay careful attention to each element of your kit.

It's also worth mentioning that the instrument you bring up first is genre-dependent. For most genres that use 'real' instruments, starting with the snare is the right choice. The snare is not as important as the kick in many electronic genres. Use your judgment to decide which element to build your drum balance around.

> TASK – Practise the bottom-up technique with a new set of stems in a new project.

Day 12 – Top-down mixing

The other option for balancing your drums is the top-down approach. As you can imagine, this is the reverse of what you previously did. The bottom-up approach is excellent for electronic styles of music. The top-down approach is ideal for tracks with acoustic drum kits. Rather than starting with your snare, you will begin with your room mikes and bring everything else down. You start here because the room mikes are the most natural representation of the drum kit. So have your room mikes at unity.

Into this blend go your overheads and any other cymbals that you have. Every time you bring a new element into your mix, you want to ask yourself, 'Am I making this sound better?' Don't add in parts just because you have them. Add them because they're contributing to a better-sounding mix. If they're not contributing positively, don't use them!

Once you have all your cymbals in, you can add your snare channels, kick, and toms. The same balancing rules apply as before: You want your snare to feel prominent as it's the foundation of your backbeat. Your kick should sit just below this, feeling complete and robust, without interfering with the bottom end of the snare drum. The toms, as before, should be sitting just below your snare.

I'll say this many times throughout this course but use your reference tracks. How you mix your drums and many other aspects of your mix will depend on the style of music you're trying to make. To get a genre-appropriate sound, comparing what you do side by side with professionally produced and released tracks is essential. This isn't cheating, and it doesn't make you less competent. All the pros do it! It keeps your ears honest and trustworthy and puts your mix decisions in context. Think about it like this: If I gave you a blank canvas and beautiful paints and brushes, you'd have everything you needed to make a beautiful painting. But if I blindfolded you, you couldn't see what you were doing! Mixing without reference tracks is the same. Using reference tracks is your way of seeing that the mix decisions you've made are good ones, rather than doing it blindly and hoping for the best.

> TASK – Practise the top-down technique with a new set of stems in a new project.

Day 13 – The clockface technique

So now you've got a balanced drum sound, the next thing you're going to do is pan it. I will cover the whole subject of panning in detail in Unit 2, but I wanted to touch on drum panning now as it's something people want to implement immediately.

There are generally two schools of thought when it comes to panning drums. Either, you pan the kit from the audience's perspective, which is the more traditional approach. Or you pan it from the drummer's perspective, which is the more modern approach. However, dependent on whether you have a right-handed or left-handed drummer, this makes no difference whatsoever! Instead, ask yourself: 'Do you want to hear your toms descending from left to right or right to left?' My personal preference is from left to right, which would be the drummer's perspective for a right-handed drummer. But then, I am a right-handed drummer, so it makes sense that I like this perspective. The most important thing here is to ensure that your overheads and rooms match however you pan your instruments in your kit. What does this mean? Well, let's say you've recorded a right-handed drummer with toms descending from left to right. But then you want your toms to be panned in the mix from the audience's perspective in the traditional way, going from right to left. This means that your individual toms will be going in contrary motion to your overheads and rooms. So, you'll need to reverse your panning on your overheads and rooms if this is the case.

12 Unit 1: Balancing a mix

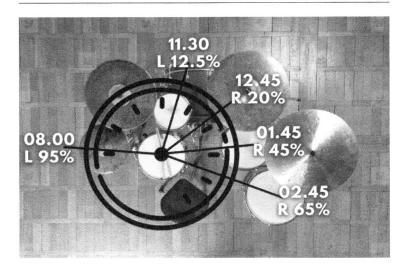

Figure 1.1 An aerial view of a drum kit with clockface overlaid to illustrate the clockface technique.

Source: Image by Sam George.

But how much should you pan things? I like to use the clockface method. In this method, you think of your kick and snare as 12 o'clock, which is dead centre in your mix. In Figure 1.1 you can see how I've identified where I'll place everything in this drum kit. I've assumed that 8 o'clock and 4 o'clock are my widest positions left and right. These are effectively 100% left and right. Then, everything else in the mix is panned according to where it is on the clock face. For example, the rack tom is at 11.30 on the clock face, which is roughly 12.5% left. The second floor is around 2.45 on the clock face, about 65% to the right. You can work out these percentages as a pan position relative to your DAW's system. Logic Pro goes from 0 – 64 left and right. Other DAWs go from 0 – 100.

Every drum kit will be different as all drummers set up their kit differently. I like this approach so much because it means that your instruments will appear in the same position as they do in your overheads. Consider it this way. Our second floor in this example will appear in this 65%-to-right position in our overheads. If we don't pan the spot mike in the same place – say we go only 30% to the right or too far at 85% to the right – it will sound like two different toms in our mix rather than one.

> TASK – Practise panning your drum channels using the clockface technique.

Day 14 – Mono compatibility

The last thing you need to do with your drums here before moving on is to check for mono compatibility. Why? For a few reasons: First, most mobile phone speakers are in mono, or at best the stereo speakers on the phone are very poor (although this is improving all the time). So, if someone plays your track out of their phone, you want it to sound good. Secondly, FM radio essentially receives mid/side transmission. If you're listening on a mono radio, the mid signal will be played out while the sides are entirely ignored. Even on a stereo radio, the radio is designed to revert to a mono transmission if the signal drops. Thirdly, some club systems are entirely mono. This is quite unusual, but commonly clubs will have a crossover where frequencies below 100Hz are in mono, and everything else is in stereo. So, the long and short of it is that mono compatibility matters! Even though you hope that everyone will listen to your music through a lovely stereo system, sometimes this is out of the listener's control.

Pull up a metering plugin on your stereo out to check your mono compatibility. I like Levels from Mastering the Mix, but you can easily use your DAW's stock metering plugin. You're looking for a positive correlation: Something above 0. The closer to +1 it goes, the more in phase it is. If it drops below 0, it means you have some phase issues. As we know already, phase cancellation happens when you have two or more audio signals moving in contrary motion that cancel each other out. In stereo, they will sound fine as one will be panned left and one right. But when you make your mix mono, they'll become part of the same signal and will therefore cancel each other out.

To hear what elements are causing you issues, you'll want to listen in mono so you can hear what's falling out of your mix. To do this, pull up a directional mixer and narrow it to 0 so that your mix is in mono. Then listen carefully to identify which mix elements you need to rework to avoid phase cancellation problems. Alternatively, lots of metering plugins also have a mono switch built in.

TASK – Insert a correlation meter on your mix buss and check your drums for mono compatibility.

Day 15 – The VU meter trick

Once you're happy with the balance of your drums, it's time to bring your bass into the mix. Bass is always challenging to gauge due to the amount of low end there is. You're looking for bass that's strong and big, but not so much that it overpowers your kick drum.

Make sure that you check your reference mixes at this point. The amount of bass that can be heard in a mix changes a lot from genre to genre. Don't be afraid to adjust other levels in your mix as you add more parts. Every time you bring a new element in, it's likely to interfere with something else slightly. Nothing is set in stone. You can make adjustments at any point.

I've got a great trick to help you get the balance between your kick and bass just right, though. Solo out your kick drum. Then adjust the trim on your VU meter until single hits are metering at -3dBVU. Then add your bass in. Trim your bass on its channel so that when the kick and bass strike together they hit 0dBVU. Nine times out of ten this will give you an outstanding balance between your kick and bass. Just don't forget to return the trim on the VU meter afterwards.

> TASK – Practise the VU meter trick on a range of different kick and bass channels.

Day 16 – Balancing vocals

It's now time to bring your vocals into your mix. It's an excellent idea to bring them in at this stage to ensure they are the focal point. If you wait until after you have your guitars, synths, or keyboards in, it can be tricky to find the right balance and can often sound like your vocals have been plonked on top of the mix rather than belonging to, and being part of, the mix. The earlier you bring in the vocals, the easier it is to make them the focal point. Your vocals should be the loudest element in your mix but shouldn't be so loud that they feel separate from the rest of your mix. They should feel pronounced but not disconnected.

In terms of what you see registering on your meters, your drums will still appear to be louder than your vocals. This is because drums are short, transient heavy sounds that come and go very quickly, whereas vocals are a longer, more sustained sound. Therefore, your drums will likely have a higher peak level, but your vocals can still feel louder in the context of your mix. This is why it's so crucial that you mix with your ears as well as your eyes.

> TASK – Study the level of the lead vocal in a range of genres. You will find that it is more prominent in some styles than in others. How will this inform where you place your vocals?

Day 17 – The mixing hierarchy

Next up, rank the rest of the elements in your track in order of importance. It's vital to work out your hierarchy of mix elements before you start to bring things in. This is because you can only really have one main thing in your mix at a time. You'll find that if you try to have two equally balanced focal points in your mix, they're more likely to fight for your attention rather than complement each other. By working out how integral an element is to your mix, you're deciding what you want your listener to be drawn to and what will play more of a supportive, contextual role. Use your reference tracks to inform your decisions here. Then, bring all these parts into your mix one at a time in the order you've specified.

Make sure you continue to check your mono compatibility, although this shouldn't be an issue at this stage as you're not going to be panning anything yet. Just remember, only one instrument can be your focal point at any single moment in time.

TASK – Analyse a few songs in different genres. Write down what you consider to be their mixing hierarchy based on the mix that you can hear.

Day 18 – A singular focal point

It can be tough to learn how to balance all the elements in your mix. I need to emphasise that there is no one-size-fits-all cheat sheet for this. How you balance your parts will vary enormously depending upon the style of music you're producing. In rock and metal, your guitars will be fundamental; in hip-hop, it'll be your kick and 808; in EDM, it'll probably be your bass; in pop, it'll be your vocal. It'll take some time to get this skill down, which is why I felt it was essential to start with the topic of balancing your mix. It really is the most important, yet most overlooked, aspect of mixing.

But here's the vital point: Balancing your mix first before applying any EQ, compression, or effects makes it much easier to address frequency and dynamic issues later. You've already created a clear vision for how you want your mix to sound. The steps I've outlined so far may seem quite time-consuming, but they're the complete opposite. Rather than setting out on your journey without looking at the map and just blindly hoping that you end up with a successful mix, you've set yourself up for success, creating a clearly defined roadmap pointing you in the direction you want to go. Just remember this: It doesn't matter what gear you've got. You can produce entirely professional, release-ready tracks with stock plugins. It's all about your ear, and training yourself to make positive, informed mix decisions based upon a clearly defined goal.

> TASK – Go through a few of your previous mixes. Write down the song structures, and then identify what the singular focal point should be for each section. Is this clear, or do you need to make some level adjustments?

Day 19 – The rule of four

The process we've just gone through is called making your static mix. It's essentially the process of balancing your mix without applying any additional techniques and without any automation. Hence, everything in your mix is static. An excellent way to think about this is by remembering the rule of three. This is sometimes just referred to as the rule of two. These are changing the level and panning. We've talked a lot about changing the level, and panning we'll cover in detail in the following unit of this course. The third part is using the polarity-inverter, which we've also talked about already.

However, I like to think of it as the rule of four, the fourth being the trim plugin, which should always be placed in the signal chain's first slot. Using the trim plugin to get your level on your channels to around 0dBVU on your VU meter, you will set yourself up to begin from a really balanced starting point. Therefore, you will reduce the amount of movement required on your faders later.

If you can nail these four skills, you will be able to make balanced mixes time and time again and will be setting yourself up to succeed.

> TASK – Practise setting up a few static mixes implementing the rule of four. You can overlook the panning for now, as we'll cover that in the next unit. You should place a gain plugin on each channel, trim referencing the VU Meter, check the polarity, and then balance your mix using the faders. Ensure this is engrained into your process.

Day 20 – The pink noise technique

There's another interesting technique for mix balancing called the pink noise technique that I thought I'd share with you. The method is this: You set up a pink noise generator within your session and set it for the desired target loudness (let's stick with 0dBVU). You then go through and solo each track one at a time and adjust your trim until it's just audible above the pink noise. In theory, once you're finished, you'll have achieved roughly the same thing: All tracks should feel nicely balanced, with no one sound jumping out too much above the rest. This should also provide you with a solid starting point and make it easier to preserve your headroom.

Every DAW will have a pink noise generator built in somewhere, so seek it out. It may be within the utility plugins, or at the very least, some synths have a pink noise generator within them.[4]

TASK – Set up a new static mix using the pink noise technique. How does this compare to a static mix made in the traditional way?

Day 21 – Loudest first

I said it earlier on, but I think it's important to emphasise. The best place to start is with your loudest or most complex section. They tend to be one and the same. If you can balance the part of your song that's busiest, where there's the most going on, the rest should fall into place without too much further head-scratching.

Let's think about this for a moment. If you're able to balance many intricate parts going on, and you can hear everything in your mix at the appropriate level, you must be setting yourself up for success. Remember, the most critical thing you're trying to achieve at this point is a balanced mix that you can then use as a springboard forward. You will use the balance you've created to compare back to at every stage in the rest of your mixing process, continually asking yourself, 'Am I making this sound better?'

This is also a good point to remind yourself of the hierarchy of mixing we touched upon earlier. Remember to bring your instruments into your mix in order of importance. Here's a good breakdown you can use: Drums, bass, lead vocals, other lead instruments/sounds, harmonic instruments, backing vocals and other melodic parts that aren't lead lines, and lastly, any contextual elements such as pads or sound effects.

TASK – Go back to a previous static mix. Pull all the faders down and rebalance it, focusing on the loudest section of the song. How does this static mix compare with the previous one you had?

Day 22 – Set a timer

It can be easy to spend hours and hours working on balancing your mix. But remember, every move you make at this stage will compromise something else in your mix. Turning up your kick will impact your snare, turning up your bass will impact your kick, turning up your lead part will impact your lead vocal, and so on. There are many other things that you'll implement later on, such as panning, EQ, compression, and automation, that will help all of these things work with each other in your mix rather than fighting. So, at this stage, try not to spend too long. If you're an indecisive person, I recommend setting yourself a timer. Start with ten minutes on the clock and see if you can get it down to five minutes with a bit of practice. This will really help to focus your mind and your ears.

> TASK – Get a timer ready! Set it to ten minutes, and then see if you can make your static mix in that time. Practise working efficiently until you can reduce this time to five minutes.

Day 23 – Leave your busses

If you're like me, and you like to have your session well organised, you'll be using busses or track stacks to route things. You may have all your drum tracks coming out of a drum buss, all your guitars out of a guitar buss, etc. At this point in the process, try to leave the faders on your busses at unity gain. If you need to make adjustments, do it on the individual channels within the busses. It's a good idea always to have the faders within a buss less than that of the buss itself. This is just good practice, and it will help in your quest to preserve headroom throughout your mixing process.

> TASK – Review your busses in a mix. Are they at unity? If not, adjust things so they are!

Day 24 – Watch your meters

Watch your meters throughout your mixing process. I can't stress this enough. You *must* keep an eye on your levels as you mix. The last thing you want to do is end up going too hot and clipping. And the more channels you have in your project, the more likely you will reach this point. So, keep an eye on it! We've already mentioned our peak level, but let's remind ourselves. On individual channels, you don't want your peak level going above, say, -6dB at the very most. For me, this is too hot. I set -10dB as my max because I don't want to exceed -6dB on my mix buss.

Peak meters are one thing, but something else entirely is LUFS. LUFS stands for Loudness Units Full Scale and is what matters when it comes to competing with professionally produced records. It can be measured in three ways: Momentary, short term, and integrated. Momentary is how loud your track is at that specific moment in time, short term is over a period of a few seconds, and integrated is over the length of everything you play through it. At this stage, keep an eye on your short term. I would recommend aiming not to exceed -18dB LUFS short term at this point.

Most DAWs will have a loudness meter built in, and if they don't, there's a great free one you can grab from YouLean. You can adjust your target level within it on some, which is very handy!

> TASK – Set up a loudness meter on your mix buss. You can use your DAW's stock meter if it has one or get a third-party option. Monitor your short-term loudness during the loudest section of your track. What numbers are you metering?

Day 25 – Split them up

Here's one last step that I advise you to do to give yourself maximum control over your mix. Split tracks up. What I mean by this is this: Say you've got the lead vocal all on one track, it's likely that there will be a lot of dynamic variety between the verse and the chorus. So, split this into two channels so that you can process them differently. Or perhaps you've got the acoustic guitar all on one track, but the verse is fingerstyle, and the chorus is with a plectrum. Basically, look for anything where there's an identifiable difference in timbre or performance and split this up onto additional tracks. In the short term, this will mean that you can make your mix more balanced at this initial stage *and* you're in control of more of your mix, but it also means that later, you have more flexibility over how you approach different elements. This approach also means that you're likely to avoid any large, clunky automation moves later, too.

> TASK – Revisit a mix. Are there places where you could split channels up? The likely candidates are vocals and guitars, but you may find opportunities on any instrument. It will depend on the arrangement.

Day 26 – FFT metering

The last thing I strongly recommend setting up in your mix process is an FFT meter. Again, most DAWs will probably have a version built in. There's also a great free one from Voxengo called SPAN that many people like. FFT stands for Fast Fourier Transform, which doesn't provide much of a clue about what it does! It essentially gives you a visual representation of all the frequencies present in your mix in real time. They're great for keeping an eye on the frequency content you have in a mix, helping you identify where you may have too much of something or not enough of another. In general, a balanced mix should have a full range of frequencies present, more or less hitting 0dB on your FFT reader. An FFT meter will help you identify if you are falling into any common traps: The smiley mix that sounds hollow, the bright mix that kills your ears, or the bass-heavy mix that can sound dull and boring.[5]

> TASK – Set up an FFT meter on your mix buss. You can use your DAW's stock meter if it has one or get a third-party option.

Day 27 – Metering checklist

Let's remind ourselves of the meters we should have set up, then. First up, you should have a VU meter calibrated to -18dBFS. Then you should have your correlation meter and your loudness meter. Often you can get these rolled into one, if you use Levels, for example. Or you can pull up individual plugins if you prefer that. After that, you want your FFT meter. Again, this may be rolled into one if you have a nice multi-meter or use a standalone plugin. After all my meters I have my referencing plugin, Reference. The only other plugins that should be present in your mix at this balancing stage are your gain plugins on your channel strips that you're using for trimming and checking for phase cancellation.

As I said at the outset, mix balancing is a simple concept, which is why it's so overlooked and, therefore, frequently, so poorly done. Spend time training your ears, training yourself to use your meters properly, and working at appropriate levels. I promise you it will make the rest of your mixing process 100 times easier. Don't be lazy. Get the basics right. They truly are the foundations for a great-sounding track.

> TASK – Run through your metering checklist. If your DAW has the option to save a channel strip as a preset, then save your mix buss setting. This will allow you to recall all your required meters quickly and easily.

Day 28 – Unit summary

So, there we go. Unit 1 is done. Let's quickly recap everything we've learned this month. We've learned about:

- Sound pressure levels and perceived loudness
- The importance of working in 24-bit
- Using reference tracks to keep your ears honest
- Gain staging the right way!
- Pre-fader metering
- Checking for phase cancellation
- Bottom-up and top-down drum mixing
- The clockface method for panning drums
- Checking mono compatibility
- The mixing hierarchy
- The rule of two, or three, or four!
- The pink noise technique
- dBFS vs dBVU vs momentary, short-term, and integrated LUFS
- FFT metering

That's a lot of ground to have covered. I promised you I'd be thorough, and I hope I haven't disappointed you. See you in a couple of days when we'll be looking at panning.

Checklist

- Make sure you're working from a clean mix
- Check your loudness at your mix position
- Check your DAW is set up to use 24-bit
- Set up your reference tracks
- Gain stage with a VU meter
- Toggle pre-fader metering
- Bottom-up or top-down drums
- Pan your drums
- Check mono compatibility
- Bring in other mix elements according to your mix hierarchy
- Do you know the rule of four?
- Have you tried the pink noise technique?
- Set a timer
- Split tracks up where necessary
- Set up your meters (correlation, loudness, FFT)

Further reading

1 Triggs, R. (2021). *What you think you know about bit-depth is probably wrong.* [online] soundguys.com. Available at www.soundguys.com/audio-bit-depth-explained-23706/ [Accessed 9 Nov. 2022].

2 Asher, J. (2022). *Why use pre-fader metering in Logic Pro X?* [online] macprovideo.com. Available at www.macprovideo.com/article/audio-software/why-use-pre-fader-metering-in-logic-pro-x [Accessed 9 Nov. 2022].
3 Hobbs, J. (2021). *What is phase cancellation? Understand and eliminate it in your audio.* [online] ledgernote.com. Available at https://ledgernote.com/columns/mixing-mastering/phase-cancellation/ [Accessed 9 Nov. 2022].
4 Bazil, E. (2014). *Mixing to a pink noise reference.* [online] soundonsound.com. Available at www.soundonsound.com/techniques/mixing-pink-noise-reference [Accessed 9 Nov. 2022].
5 NTI Audio. (2014). *Fast Fourier Transformation FFT – basics.* [online] Nti-audio.com. Available at www.nti-audio.com/en/support/know-how/fast-fourier-transform-fft#:~:text=The%20%22Fast%20Fourier%20Transform%22%20(,frequency%20information%20about%20the%20signal [Accessed 9 Nov. 2022].

# Unit 2

Panning a mix

Day 1 – Panning a mix

Welcome to Unit 2 in my music-production course. In this unit, we will cover panning. Panning is one of the Big Four when mixing a track. In the last unit, we focused on the most overlooked element – balancing a mix. Panning is the next most overlooked skill to learn. Once you have a balanced mix as your starting point, panning is your first opportunity to create space between elements. Rather than just having all your channels equally positioned between your left and right monitors or headphones, you can place things further to the left or right, giving them their unique position in the stereo field.

But how do you do this in an informed way, with strategy and method, rather than fiddling and guessing? That's what I'm going to cover in this unit. No longer will panning be the second most overlooked tool in the music-production toolbox! Let's do this.

Day 2 – Know your pan pots

Before you pan anything in your mix, you need to understand this: There is more than one type of pan knob. Different DAWs present their channels in different ways and therefore present their pan pots differently. I'll use Logic Pro as my example, but you'll need to investigate the arrangement for your DAW of choice. In Logic Pro, there are three types of pan pot. Many producers don't know this, let alone understand the difference. In Logic, you'll be able to change how your pan knob behaves by right-clicking on the pan pot on the channel strip. You can then select one of the three options: Balance, stereo pan, and binaural.

By default, your pan pot will be set up as a balance knob. This means that your signal's level will be adjusted in your left and right monitors according to how far you pan to the left or right. For mono tracks this is perfect. Given that mono tracks contain just a single sound source with no stereo information, using a balance knob to adjust how much of it you hear in the left or right speaker is ideal.

However, for a stereo sound source, this isn't so helpful. Stereo tracks have different information on the left and right channels. Using a balance knob on a stereo track will only adjust the left and right channel's level rather than changing the whole sound source's position to the left or right. So, if you pan a stereo track hard to the left with a balance knob, you will effectively just hear the left channel of the sound and none of the right.

Instead, you want to use a stereo pan pot on stereo tracks. This type of control will change the stereo channel's absolute position between your monitors, not just adjust the left and right channel's levels. This means you'll still hear your entire stereo instrument, only in the stereo field position you want.

As a general rule, on mono tracks, use the default pan knob, a balance knob; for stereo tracks, use a stereo pan knob.

The third option is a binaural pan control, which gets pretty tricky, and I'm not going to be covering it in this unit. In a nutshell, it's a method of emulating human hearing by allowing you to position a sound source in front, behind, above, below, or to the left or right of the listening position.

> TASK – Explore the pan pots in your DAW. Ensure you know what pan pots you have available to you and how you change between the different types.

Day 3 – From whose perspective are you mixing?

From whose perspective are you mixing? This is an important question to ask yourself from the off before reaching for the pan pot. I touched upon it in the last unit when talking about drums, but that same principle applies to your whole arrangement. This question is particularly relevant when mixing live instruments rather than synthesised ones, although the same concepts can be applied to an electronic arrangement.

The question is this: Do you want to pan your mix from the audience's perspective, as if you're at a gig, watching the music being performed? Or do you want to feel like you're in it, like you're on stage with the musicians physically within the performance? This question is crucial if you're working with a live recording rather than a multitracked one, where there may be elements of bleed between the microphones in your recording. In all scenarios, however, it's good to decide on the perspective. It will inform your choices on tom direction, percussion, guitars, keys, horns, BVs, and all sorts of other things.

> TASK – Listen to some records in contrasting genres. Some should be studio recordings, some live recordings. Pay attention to the panning of the tracks. From whose perspective are you listening?

Day 4 – Kick and snare

So, let's talk about drum panning. You're likely to have both mono and stereo tracks within your drum kit. For example, your kick, snare, and toms will be mono, but your overheads and rooms may well be in stereo. So, make sure you've got the relevant pan pots set up, balance knobs on your mono tracks and stereo pans on your stereo tracks. I'm assuming here that your drum kit is already balanced and that you've solved any phase issues.

Now let's focus on the kick and snare. 99 times out of 100 your kick and snare will stay perfectly centred in your mix, straight down the middle. They are the fundamental components of your backbeat, so you usually want them to provide security by being right down the centre. But if you listen to early Beatles records, for example, you'll hear that this isn't always the case.

Whether mixing live drums or sequenced ones, it's very common to layer samples. This may be via drum replacement or triggering on a live kit, or by adding a range of samples to create depth in electronic music. Often, samples that you layer will have been recorded in stereo. You can leave them in stereo to help create width in your kicks and snares. However, it's a good idea to keep an eye on the amount of low-frequency content you're adding and try to avoid adding too much low-frequency content to the sides of your mix. To prevent this, you can either use a high-pass filter, use some mid/side EQ to remove the low end from the sides or use a stereo imager to narrow the stereo image, bringing it closer to the centre and away from the sides of your mix. All these EQ concepts will be covered in more detail later in the course.

Particularly in electronic music, where you are layering snare samples, you may wish to place different sounds in slightly different positions in the stereo field to widen your snare sound. You may utilise various combinations of samples between different sections of your song. There are no hard rules for this. My advice is this: Ensure you stay balanced throughout, with an even amount of snare on each side of your mix.

> TASK – Pay attention to the kick and snare in various records. Listen to some acoustic tracks, some electronic, some old, some new. What do you notice?

Day 5 – Overheads and rooms

Panning your overheads and room mikes is pretty simple. Often, you'll have stereo overheads and stereo room mikes, and sometimes you'll have a mono overhead and mono room mike too. Assuming you've already decided whether you are mixing from the audience's or the drummer's perspective, you will pan your stereo tracks accordingly. If your stereo tracks are interleaved, meaning they are both on one stereo track, then you don't need to adjust anything concerning the panning, except for checking that you have a stereo pan pot set up. If your stereo overheads and rooms are delivered as two mono tracks that are not interleaved, then you have a decision to make. How wide do you want to pan these mono tracks to create your stereo image?

Generally, shooting for about 75% to the left and right is a good starting point. You probably don't want to go hard left and right, as this could well start to create some phasing issues. My big tip here is that you can use the width of your stereo overheads and rooms in combination with your mono overhead and room (if you have them) to create great variety in your mix. Using more mono tracks and fewer stereo tracks in verses and vice versa in choruses will make a nice opening-up feeling, providing a much-needed lift into your hook. If you don't have mono overheads and rooms, you can create the same effect by narrowing your stereo tracks' pan position in your verses and widening them in your choruses. If you're working from interleaved stereo tracks, you can use a stereo imager to narrow your verses. This is a topic I'll cover in detail later in the course, but it's a great tip to start playing around with now.

This concept applies to all genres. To create a subtle lift into your chorus or drop, widen your percussion so it opens out into your chorus and has more space.

It's essential that you know how the overhead mikes were positioned for the recording. If they were evenly spaced, then you're fine to pan your channels in equal proportion. However, if a technique such as the Glyn Johns method was used for tracking, you need to know about it! I recommend researching this arrangement if you don't know about it already. You can hear it on classic records from the Rolling Stones, the Who, and Led Zeppelin, amongst others.[1]

> TASK – Research the Glyn Johns Microphone Technique

Day 6 – Hi-hats

Panning hi-hats is a hot topic of debate that people argue over for hours. My opinion on it is simple. If you're working with a real drum kit played by a human, then the hi-hat will be panned to whatever position it was on the drum kit. Often this will be around 60–70% to one side – whichever side it was on the kit. However, don't feel that you *must* use it. You may find that you have enough hi-hat present in your overheads and that an additional hi-hat channel is overkill. It will depend on the part, the performance, and the recording. So, use your ears and producer's intuition.

In electronic music, however, the rulebook is far less precise. Sometimes the hi-hats will stay very close to, if not straight down, the centre of your mix. Sometimes you'll be working with a stereo hi-hat sample. From time to time, you'll want to hard-pan your hi-hat. Frequently, you'll be using multiple hi-hat samples in the same project. Occasionally you'll use auto-panning so that the hats move around the stereo field. So, to put it simply, you can do what you want with your hi-hats in electronic music. In this case, I would advise using the guidance I just gave you: Think about having narrower verses and wider choruses to create movement in your mix. Keep your verse hats narrow and close to the centre and widen them out in your hooks with more movement and panning.

> TASK – Compare the hi-hat positions between acoustic and electronic records.

Day 7 – Toms

Panning toms on a live drum kit is reasonably straightforward. You want to place them in the same position in the drum kit as they appear in your overheads. Think back to the clock face technique from Unit 1. That's your best starting point for live drums. As a general rule, I wouldn't go wider than about 75% to the left or right for toms. But this is dependent on the size of the kit. How you pan Travis Barker's toms will be very different from what you do with Mike Portnoy's!

For electronic toms, as with your hi-hats, there is no specific set of rules to follow. In the case of electronic music, do what is best for the song. If you want a crazy super-wide tom fill, go for it. But always do what is best for the music. Generally, you want each mix element to have its own space in your stereo field. So, don't place your mad tom fill directly on top of your hats, shakers, or congas. Position each element in its unique spot so that it has a place to poke through the mix without interfering with another aspect.

> TASK – Practise positioning your toms in the same position as they appear in your overheads. Adjust the width of your overheads and adjust your toms accordingly.

Day 8 – Bass

Panning bass is simple: Don't pan it! Your electric bass, along with your kick and snare, is what will root your track, so keep it centred. However, sometimes you may wish to add some subtle roomy width to create depth to your bass and make it feel a little more 3D. You can do this by creating a reverb send, but keep it tastefully low in the mix. Generally, I will let this come forward in sections of my track where the texture is thinner – most likely the verses, where there is more space in the arrangement for the bass room to be heard. But in the fuller sections of the track, you can back this off, allowing the bass to entirely focus back down the centre of your mix whilst the song is at its busiest.

Synth bass is a slightly different animal. For thick, subby basses with loads of low-end and very little high-frequency content, you'll want to keep these up the centre of your mix. For a gritty, biting bass line with a lot more melody that is used more like a riff or a bass lead, you'll often want this to have more width. Especially if you're using an instrument in stereo, you'll want to keep it this way. In this instance, I suggest you keep everything below 100–120Hz in mono, and anything above this can be in stereo. You can achieve this in a couple of ways. Either use a mid/side EQ, low-pass everything in the mid-channel below 100–120Hz, and high-pass everything in the side channel above 100–120Hz. This method can feel a bit crude. Alternatively, duplicate your track, so you have the same content twice. Low-pass one at 100–120Hz and use a stereo imager to make it mono. High-pass the other at 100–120Hz for your stereo content. You can then process both elements of your bass sound independently and buss them back together to give you maximum control over your sound.

> TASK – Practise both bass techniques. Firstly, set up a reverb send and practise automating the send amount so it moves forwards and backwards in the mix dependent on the section. Secondly, practise separating your bass signal into frequency bands using filters on duplicate channels.

Day 9 – Guitars

Let's talk about panning guitars. The first vital thing to note is that guitars and voices share a lot of similar frequency content. So, if you put them in the same position in your stereo field, they will end up fighting each other. The loser will be your mix. Therefore, I recommend you aim to put your guitars in a different position than any vocal in your mix.

Double-tracking guitar parts, particularly rhythm guitar parts, is a widespread tactic for making guitars sound thick and full. I think these should almost always be panned 100% left and right, ensuring they are as far away from the lead vocal as possible. As I've alluded to, you can sometimes narrow these a little bit in the verses to maximise the widening, opening-out feel in the chorus. However, I would still recommend keeping your double-tracked guitars close to the extremes of your stereo field. There are other ways to achieve weight in your chorus, which I'll cover later in this course.

For single-tracked guitar parts, you need to train your ear to balance your mix's different components. Think of your mix's left and right sides like a set of scales. You want the two sides to balance each other out. So, if you have a single guitar panned a quarter to the left, you may balance it with a horn a quarter to the right. Or a guitar halfway right may be balanced with a synth halfway left.

Lead guitar parts are generally close to the centre of your mix, as they effectively behave as melodic lines. Double-tracked leads may be panned 10% or so left and right, but you can play with this a bit to find the right balance. There are other tricks you can use to get it out of the way of your vocal.

> TASK – Listen to some guitar-based music. Analyse the panning relationship between lead vocal, lead guitar, and rhythm guitar. What do you notice?

Day 10 – Synths and keys

Synths and keys are tricky to cover with a specific set of rules as there's so much variety here. My big tip is to consider whether something *needs* to be in stereo or not. Almost always, synth and keys instruments will come in stereo and will automatically fill a lot of space in your mix. This can be useful for things like pads that are designed to fill out your mix without being a focal point. But for more melodic synth parts, this can be counterproductive. You may want to consider narrowing the stereo imaging of particular sounds in your mix or making them entirely mono. This will help you manage your stereo field better, keeping things in more clearly defined positions and helping your ear navigate your mix.

In general, pay close attention to your midrange between 250Hz to 2.5kHz. Things that sound great in stereo in isolation will often sound less useful in the context of a mix and will benefit from having a more defined position. Similar to how you treated your individual guitar channels, melodic synths need to be nicely balanced with equal weight on both sides of your mix.

For stereo tracks such as piano, organ, and pads, if you find that they get a little boring when they are placed in one static position in your mix, you can try utilising an auto-panner to add some movement to them. CableGuys' Pancake is a great free plugin. Or Soundtoys' Panman is a brilliant premium option.

> TASK – Review the keys and synths in some of your previous tracks. Assess whether some of these parts may have been better off narrower in the mix or entirely mono.

Day 11 – Horns, wind, and strings

Horns, wind, and strings generally fall into one of two categories. You'll either have full orchestral sections or smaller groups of instruments – perhaps a handful of horns or a couple of strings. The approach to each is quite different.

For full orchestral sections, you will usually position them as they would appear if they were sitting in a full orchestra.[2] You want to look at an orchestral map to do this. Suppose you're using a premium orchestral sample library from the likes of Spitfire Audio. In that case, the libraries come with panning built into the instrument, depending on the combination of mikes you choose to use inside the device. However, this doesn't apply to all orchestral libraries, so it's good to have a guide at hand. You can employ the clockface method again here as you did with your drums. Take note, though, that it will only work effectively in your mix if you're utilising a whole orchestral family at a time. If you just have violins but no cellos, you'll end up overloading the left side of your mix and vice versa.

If you have a smaller horn or string section in your track, lay them out as they appeared in the recording or as they would appear on stage. As with other elements in your mix, think about whether you're panning from the audience's or the performer's perspective. If you have just a smattering of solo wind or strings, think about how their timbre will interact with other elements of your mix and try to keep them apart. For example, a single violin or viola is best kept away from an acoustic guitar. A single trumpet is best kept away from your hi-hats and so on.

> TASK – Study some pictures of full orchestras. Identify where the different sections are located. Create your own panning template based on this.

Day 12 – Vocals

Always keep your lead vocal centred in your mix. I pretty much never stray from this. If I have a single lead vocal in a track, it'll be kept straight up the middle. I may stray from this if I have a double-tracked lead vocal, for example. Then I may pan each one slightly off centre. Similarly, I may pan ad-libs off centre. But they'll still be very close to the centre of the mix.

The rulebook for backing vocals is much less easily defined. Modern productions often have layers and layers of backing vocals to help thicken the texture and give a lush and rich production. I love to find pairs of backing vocals that are similarly weighted and pan them hard left and right to create a super-wide feel to the BVs that won't interfere with the lead vocal. Again, you can alter the positioning between verse and chorus to make your mix open out into the chorus or drop.

If you have fewer BVs that can't so easily be hard-panned and balanced in this way, then again, you should look for a unique spot in your stereo field to position them. You can place a BV anywhere in your stereo field. Keep in mind the main tips we've covered already: Don't put it directly on top of something else in your mix and keep things evenly balanced between your left and right sides.

> TASK – Pay attention to the backing vocals across a range of different genres. Do you notice any trends regarding the positioning of BVs that are genre-specific?

Day 13 – The LCR technique

Knowing where to pan things initially can be a very daunting task. Having 128 potential positions (as you do in Logic Pro) to place something in is a lot of choice. The LCR panning technique is a wonderfully simple yet incredibly effective method for getting a mix that works quickly, using just three positions: Left, centre, and right. Using this method, you choose to place things either down the centre of your mix or hard to the left or right. Using this method, you don't have to worry about being precise. Just place things in one of the three positions and move on. Even if you intend to use more precise panning positions further down the line, it's a great place to start, from which you can then further define where you want things to appear. But it's instrumental in ascertaining if your mix will be balanced on the left and right sides or if you're overloading one side or the other. It's advantageous when you're working on a large project with many channels. Using the LCR method is excellent for getting your mix up and running in a short time as it forces you to make decisions quickly, assessing their impact immediately. Think of it like this: When painting, instead of deciding if you want to paint something navy, baby, or sky blue, determine if you wish to paint it blue at all or if, in fact, it would be better off red. This is the LCR panning method in a nutshell.[3]

> TASK – Using the multitrack stems for a new project, set it up using the LCR technique. How quickly can you balance your stereo field?

Day 14 – Low-frequency panning

Where you pan your low-frequency content is an essential choice. As a rule, the lower the frequency content is, the closer to the centre of your mix you'll want it to be. Therefore, your kick and bass will stay in the centre of your mix because they're the lowest sounds in your mix. Keeping low frequencies in the middle of your mix gives you a solid core to work around and gives you good focus. So generally, anything below around 120Hz should be centred.

Doing this is more beneficial than just giving you a solid core. It will also help your mix translate better on different sound systems. Anything that makes a stereo system mono, such as a Bluetooth speaker or a club sound system, will reduce the overall power of panned bass frequencies. So, if your bass sound is coming from a stereo instrument, which is the case with most VSTs these days, it's a good idea to make this mono. I would recommend routing your bass instruments out of a mono buss. This can sometimes be a little too crude, however, as you may well have sub and bass frequencies in your sound, as well as more crunchy or distorted midrange, which you might want to keep in stereo. In this case, I would suggest splitting your signal into two channels. So, duplicate your bass channel so you have it twice. You can then high-pass one channel and low-pass the other somewhere between 100–120Hz with a gentler slope on the filters so that the crossover point isn't too crude. You can focus the stereo image of your low channel or make it entirely mono whilst processing the stereo image of your upper channel independently, retaining some or all of its stereo image.

> TASK – Go back over some of your previous mixes. Focus the low end of your kick and bass channels below 120Hz into the centre of your mix. How does this affect how the song feels?

Day 15 – Double-tracked guitars

I mentioned panning double-tracked guitars earlier, but I wanted to expand on it a little more here. Generally, if I have a double-tracked rhythm guitar part in my mix, I will pan one hard left and the other hard right. To clarify, this technique only works when you're working with two separate takes. It doesn't work if you just duplicate one part and pan it. This is because, if you reproduce a sound, you will merely be doubling its amplitude. It won't do anything to enrich or fill out your sound; it will just be louder. Working with two different takes, you benefit from the subtle variation in each performance, which gives you the fullness and richness that is so desirable from a double-tracked part. The same concept applies to anything else in your mix, most commonly vocals.

If you're working with two takes that are almost identical in all but performance, you can pan them hard to the left and right, and they'll sound great. If your two tracks are similar but with subtle variations in the musical content, you may wish to consider panning them wide but not completely hard so that there is still some crossover in the middle of your mix. This will help the parts blend together nicely and take attention away from each performance's subtle variations.

> TASK – Experiment with the width of double-tracked guitars in a mix. What are the pluses and negatives of having them closer or further away from the centre of the mix?

Day 16 – Complementary panning

Complementary panning is also something I mentioned earlier, but it's worth emphasising at this point. At all times in your arrangement, you should aim to have a balanced stereo field. This means you have an equal number of sounds on your mix's left and right sides. Further, as well as having similar numbers of sounds on the left and right sides, they should also even out in terms of amplitude and position in the stereo field. But they should also even out in terms of timbre, too.

An excellent way to achieve this is to match up and pair off channels in your mix. For example, you could offset a lead guitar part against a keyboard track. Or a horn line against an acoustic guitar. Or a plucked lead with an arpeggiator. You should aim to match things that share similar frequency content and place them opposite each other in your mix.

What this means in real terms is that you shouldn't necessarily expect to position something in one place and leave it there for your whole track. As elements in your arrangement change, how you balance things in your stereo field will likely need to adapt too. It's improbable that all the tracks will be playing in your song all the way through. As parts come in and out, you'll need to adjust your stereo balance to stop your mix from toppling to the left or the right.

Working out how to keep your stereo field balanced at this stage immediately after you have set up your static mix will stand you in excellent stead moving forward. Separation in the stereo field is a massively overlooked component of a mix and is such a helpful tool when planned out carefully and strategically from the outset. You should use a balance meter, which will indicate whether your mix is nicely balanced or not.

> TASK – Review the panning in some of your previous mixes. Are they evenly weighted throughout? Are things appropriately paired off and balanced in the left and right sides?

Day 17 – Snare on- or off-centre?

I know I said earlier that the snare should always be straight down the centre of the mix, and I stand by this. This is my general preference. But occasionally, you may wish to experiment with putting it slightly off centre. A centred snare can be super punchy, which is often what you want, but not always. Panning it slightly to one side can draw the listener's attention to the lead vocal or kick drum, which can be desirable sometimes. Generally, I wouldn't pan a snare further than 20% left or right.

Something to be aware of concerning your snare and where you pan it doesn't involve your snare channels directly, but the other elements in your drum kit. If you're working with a live recording, you will most likely have bleed from your snare coming down your hi-hats or toms' channels. The amount of bleed you have will dictate how far you can pan these other components of your kit. If you have a lot of bleed down your hi-hats, for example, panning it fairly wide in either direction will harm how your snare sounds, making it sound skewed one way or the other.

Another consideration is that depending upon the drum kit set up in the first instance and the overhead mikes' positioning, your snare may appear slightly more prominent in either the left- or right-hand overhead. If you pan your snare or any other element of your kit, for that matter, you can end up with something quite unnatural sounding in a different position to where it appears in your overheads. As part of the same consideration here, where the snare appears in your overheads may dictate how wide you can pan them.

> TASK – Pay careful attention to your overheads. Is your snare evenly balanced between the left and right sides, or does it appear slightly more in one channel?

Day 18 – Narrow verses, wide choruses

Producers around the world strive to make their chorus or drop feel big. But how do you achieve this? The simple truth is that you need to make a difference in your production between the chorus and whatever section comes before it. This is reasonably easy if you have a large arrangement with lots of elements to play around with. The more different parts you have, the broader the range of timbres and dynamics you have to fiddle with. But what if you only have a four-piece band? Or even less, a singer-songwriter arrangement? Then you need some tricks up your sleeve.

Concerning panning, this comes in automating the width of the panning from verse to chorus, implying widening and opening out into the chorus. I would recommend a 15 to 20% differential between these sections. You don't want the difference between sections to be so significant that it obviously pokes out of the mix. You want it to be subtle, so the listener feels it without necessarily noticing it.

But remember, you can do this on mono *and* stereo tracks. If you have double-tracked rhythm guitars panned hard left and right in your choruses, then narrow them by 15 to 20% for your verses. If you have a stereo piano or acoustic guitar, use a stereo imager to narrow the stereo image by this amount for your verses.

> TASK – Practise subtly automating the width of your mix between verse and chorus. Can you achieve a widening feeling without it being too noticeable?

Day 19 – Automating movement

Now that we're in the frame of mind of things moving within our stereo field throughout the mix, let's develop this further. For an exciting mix with subtle variety throughout without being obvious, look for elements that you could add movement to that won't be too distracting.

What are we looking for here, though? Think stereo instruments that aren't a primary feature. So, pads or keys parts that aren't focal points. Try using an auto-panning plugin on them to add a little bit of subtle movement that will keep energy in your mix. Or look for mono instruments that are panned reasonably close to the centre. Consider whether you could swap parts from left to right from verse one to two. Or could the backing vocal in the chorus move position rather than always being in the same place?

With all these things, you're not looking to do something obvious that will smack the listener in the face. You're looking to add subtle interest and character. We're not talking about automating every element in your mix – far from it. But a couple of moving parts will add an extra layer of secret spice that your listeners will love, even though they won't necessarily know it's there.

> TASK – Revisit an old mix. Find one or two elements within that mix that you can subtly alter the position of to add some added interest.

Day 20 – Checking mono compatibility

Always keep an eye on your mono compatibility. When you're adding moving parts to your stereo image, it can be easy to get carried away and lose sight of your mono compatibility. I would recommend having a balance and correlation meter up to keep an eye on throughout this process. You're looking to keep an eye on two things here. Firstly, you want to ensure that the correlation stays in the positive. The closer to +1 you are, the better. But if you drop below 0, you have significant mono compatibility issues. The second thing is to check the overall balance of your mix between the left and right speakers. This is like balancing your scales, ensuring that you have equal weight on both sides. You'll want to check your whole song for both things. As the instruments and positions of things change, their interaction and weighting will also change. So, play your whole track through, looking at these two meters and make any adjustments as necessary. It seems a bit fussy, but I promise that you'll notice a massive difference in the end when you have beautifully balanced mixes that will translate on any sound system. And the more experience you gather in this area, the less you'll rely on the meters and the more you'll trust your own intuition.

> TASK – Make sure you have both a balance and a correlation meter that you like. Your DAW may have them as stock, or you may prefer to find a third-party option.

Day 21 – Consider the club

Consider the club. Especially if you make electronic music. Club systems range from fully mono to mono bass with a crossover somewhere around 100–120Hz, and occasionally fully stereo. The mono options are much more common. The wider your mix is, the less likely it is to translate well when you play it in mono. This particularly applies to elements you may have double-tracked wide or hard left and right that are likely to cause phase cancellation issues when in mono. Skrillex is an excellent example of a producer who keeps things narrow.

So, if you produce electronic music, here are some tips: Keep your bass and sub entirely in mono. Anything below 100Hz should be straight up the centre of your mix, with nothing on your sides. This will give you a prime focus. I like to have a crossover that gradually introduces some low-mid frequencies into the sides. This will roll up to somewhere between 120–150Hz. But after this, I'll keep things closer to the centre rather than panning things wide or hard. Generally speaking, I'll avoid double-tracking things or wide stereo images on sounds. This isn't to say you should avoid a wide-sounding mix. You still want it to *sound* wide, but it needs to translate in mono. So, you'll want to check your mix for mono compatibility often.

> TASK – Arrange the panning of a mix considering the club. Ensure your bass is fully mono and the rest of your elements are kept close to centre. Does this cause you any new problems you haven't encountered before?

Day 22 – Crosstalk

Check in headphones to ensure your mix doesn't sound too disjointed or unbalanced. Lots of your listeners will listen on headphones at some point. And you want your music to sound great across all listening experiences. Listening on headphones and through monitors is a very different listening experience. This is because headphones don't have any crosstalk. Crosstalk happens when information from the right monitor reaches the left ear and vice versa. When listening through headphones, this doesn't happen, so you hear just the right side in your right ear and vice versa. This makes for reasonably contrasting listening experiences. Something that sounds great in headphones may not sound so effective through monitors.

Conversely, lots of home producers may produce almost exclusively on headphones. You should try to avoid this for precisely this reason. You may think you have a fantastic-sounding mix through your headphones, but it probably won't sound the same through monitors.

The golden rule here is to alternate frequently between monitors and headphones. Even better is to switch between different monitors and different headphones, although this is a luxury many home producers cannot afford. The more contrasting devices you audition your mix through, the better.

> TASK – Make sure you have more than one listening device in your setup. As a minimum, you should have one pair of monitors and one pair of headphones.

Day 23 – Don't overload one side

Be cautious when panning rhythmic elements in your mix. As well as keeping each side of your mix well balanced in terms of frequencies, you also want to consider rhythmic elements when panning. For example, your hi-hats and acoustic guitar will likely play a similar rhythmic pattern. Probably a 1/8th or 1/16th note motif. Panning these on opposite sides of the mix will balance each side's high-end timbre and keep each side rhythmically balanced. If you pan many rhythmic elements to one side of your mix, it will become very distracting and make your stereo field feel quite unbalanced. So, similar to when you considered pairing off single instruments against each other based upon their timbral quality, you should also try to pair off rhythmic elements against each other too. In general, if you don't have something to balance against something else, keep it closer to the centre of your mix so that you don't become too heavily weighted on the left or the right.

> TASK – Review a previous mix. Pay close attention to any parts that are playing rhythmic patterns. Are they balanced evenly, or are they spread too wide? Adjust your mix accordingly.

Day 24 – Recreating a vintage vibe

Recreating a vintage vibe throws out all the rules we've discussed so far in this unit. To get that genuinely retro vibe, you may want to consider panning your drums to the right and your bass all the way to the left. Or perhaps you'll just settle for a mono drum recording. Are you looking to recreate Glyn Johns' classic late-'60s three-drum mike technique? Maybe you're after Phil Spector's legendary 'Wall of Sound'? Perhaps you're looking for the simplicity of the Motown sound? Maybe you're keen to play around with mono reverbs and their placement in the stereo field? There are so many variables to creating a vintage sound through your panning that it would be impossible to sum them up in one short chapter.

The best thing to do here is to study the records you're trying to nod towards. There's no set of rules, as many engineers have experimented with many different techniques over the years. Listen carefully to your reference tracks and extract the parts you like.

> TASK – Find a vintage record that you like which utilises some unusual panning. Attempt to replicate the positioning in your mix.

Day 25 – Less is more

When it comes to panning, less is often more. The temptation is to get carried away, panning lots of elements wide, adding in lots of complex movement via automation, and generally making things very busy. But this doesn't always lead to a wide-sounding mix. Sometimes, the broadest sounding mixes come from just panning a few exciting elements whilst maintaining a robust and balanced centre. This has the knock-on benefit of helping your mix translate when in mono.

To assist with this point of view, consider having just one wide stereo element in your mix. This could be double-tracked guitars, drum overheads, or a stereo piano. Then make everything else sit carefully around the centre of your mix, setting your levels with precision and creating separation with EQ. This can often translate into a compelling mix, which can be much less confusing or disorientating than something with a lot of movement.

> TASK – Review a previous mix. Do you have multiple elements panned out wide? Refocus this mix so you only have one wide part with everything else carefully arranged around the centre.

Day 26 – Orchestral panning 1

What about panning an orchestra? More and more often, producers and composers are writing 'hybrid' music, with orchestral elements crossing over into popular and vice versa. But most modern producers don't have a traditional music education or know much at all about a real orchestra.

To cut a long story short, where musicians sit in an orchestra is quite a specific art, which has been developed and established over hundreds of years. So, unless you have good reasons to meddle with it, you shouldn't. Instead, your task should be to try and reproduce the natural orchestral positions as best you can.

To do this, you need to know where instruments are positioned in the orchestra. So, grab yourself a diagram. On inspection, you will notice that sections with similar ranges or timbres are positioned opposite each other around the conductor's position. For example, the violins are to the left, and the trumpets are to the right. The cellos are to the right, the horns to the left. You'll also notice that the loudest instruments (the percussion and brass) are positioned at the back, and the quieter instruments (the woodwind and strings) are towards the front. Positioning instruments 'correctly' in your stereo field will make your track sound convincing, which is undoubtedly what you're after.[4]

> TASK – Watch some orchestral performances on YouTube. Compare the orchestral positioning with the map you obtained earlier. Does everything correlate as it should, or are there some variations?

Day 27 – Orchestral panning 2

Before telling you how much to pan your sections, it's important to point out that many high-quality orchestral sample libraries already have panning built into the instrument. For example, I use Spitfire Audio's Symphonic libraries a lot, and they already have panning built in. So, check this with whatever you're using first and foremost. Bear in mind that it's highly likely your sample library will be in stereo. The stereo information built into the library is critical with orchestral samples as it will offer you the space required to create depth in your orchestra. So, make sure you keep your tracks in stereo.

With that said, if you want to position your orchestral instruments manually, here's where I'd place them: For the strings, first violins halfway left, second violins less than halfway left, close to the first but they should be differentiable. Violas are centred or slightly right, celli are less than halfway right, and basses are halfway right. For woodwind, flutes and clarinets should be slightly left and oboes and bassoons slightly right. For your brass, trumpets should be about 1/3 right, horns 1/3 left, trombones and tuba halfway right. For your percussion, your timpani and the bass drum should be centred or slightly left to separate them from the basses and tuba.

> TASK – Download the multitrack stems for an orchestral recording. For the sake of practice, narrow every track to mono and then manually position everything correctly in your orchestra. A/B this with the original stereo channels. Are they comparable?

Day 28 – Unit summary

So, there we go. Unit 2 is done! Let's quickly recap everything we've covered. We've looked at:

- Knowing your pan pots
- Deciding whose position to mix from
- How to pan all sorts of different instruments
- LCR mixing
- Low-frequency panning
- Double-tracked guitars
- Complimentary panning
- Panning automation for wider choruses and drops
- Checking mono compatibility
- Crosstalk
- Not overloading one side
- Orchestral panning

As you've discovered by now, if you didn't know already, panning is way more involved than most people think. It is a massive part of making a mix work, ensuring you can hear everything in your mix clearly, and, when used intelligently, can make your mix shine. When used poorly, it can pretty much ruin your mix! So, it's well worth spending time practising this skill and getting good at it.

In the next unit, we're moving on to number three of the Big Four: EQ. I know this is a topic that many of you will want to explore in a lot of detail, so make sure your pencil is sharp.

Checklist

- Make sure you have the right types of pan pot set up
- Ensure you've decided on your mixing perspective
- Check your drum positions correlate with the overheads and rooms
- Focus your low-frequency content
- Employ complementary panning. Are your left and right sides balanced in terms of numbers? Amplitude? Timbre? Rhythmic content?

- Automate your width to widen your chorus/drop
- One or two moving parts for interest
- Check mono compatibility
- Consider the club – if it's a club mix, make it narrower overall
- Monitor using more than one method
- Sense check: Are you overdoing it?

Further reading

1 DRUM! (2021). *What is Glyn Johns technique?* [online] drummagazine.com. Available at https://drummagazine.com/glyn-johns-technique/ [Accessed 9 Nov. 2022].
2 Shaw Roberts, M. (2019). *Why are orchestras arranged the way they are?* [online] classicfm.com. Available at www.classicfm.com/discover-music/orchestra-layout-explained/ [Accessed 9 Nov. 2022].
3 Houghton, M. (2021). *LCR panning pros and cons.* [online] soundonsound.com. Available at www.soundonsound.com/techniques/lcr-panning-pros-and-cons [Accessed 9 Nov. 2022].
4 Westlund, M. (2018). *Essential tips for orchestral positioning and mix panning.* [online] Flypaper.soundfly.com. Available at https://flypaper.soundfly.com/produce/orchestral-positioning-mix-panning/ [Accessed 9 Nov. 2022].

# Unit 3
EQ

Day 1 – Applying EQ

We've arrived at one of the big ones: EQ. Before we get into what EQ is, why it's so helpful, and how to use it, it's essential to clarify that you shouldn't jump straight in at this point. If you skipped through the first two units to get to the meaty stuff, go back and do those first. EQ is a beautiful tool and will help shape a lot of your mix's clarity and character. The way to get the most out of your EQ is to balance and pan your mix thoughtfully first. If your mix isn't well balanced, then any EQ alterations you make will be from an uninformed position. So, you won't be able to gauge properly whether the EQ changes you're making are genuinely making a positive contribution to your mix or not.

Similarly, if you EQ before you pan, you may end up EQ'ing things that don't need to be touched. You may create the tonal separation you need between instruments by positioning them differently in your stereo field. If you EQ before panning, you may try to separate things too aggressively when in reality a bit of panning might get you most of the way there. So, don't cut corners in getting to this point. Set yourself up for success by doing the first two stages in your mix process: Balance your mix carefully and pan it intelligently.

What are we going to cover in this unit? We'll start by ensuring we understand precisely what EQ is and how it works. We'll talk about different types of equalisers and their strengths and weaknesses. We'll cover a range of varying EQ techniques and applications and look at common mistakes made with EQ, and I'll share some top tips for EQ'ing in an informed way.

Day 2 – Fundamentals and overtones

What is EQ? EQ is short for equaliser. An equaliser is a device that allows you to adjust the volume of individual frequencies of a sound. So rather than turning the whole sound up or down, you can just turn a specific frequency of the sound up or down, giving you more control over that sound. Any sound comprises of two parts: The fundamental and the overtones. The

44 Unit 3: EQ

Figure 3.1 A spectrum analyser indicating the location of the fundamental frequency.

Source: Image by Sam George.

fundamental is, if you like, the actual pitch of the sound. It will be the lowest part of the sound source, the initial vibration. Looking at the graphic display in Figure 3.1, you can see the fundamental sticking out. Overtones are subsequent vibrations that have their origin in the fundamental. They are generated based on something called the Harmonic Overtone Series. The first overtone will be an octave higher than the fundamental. The second will be a fifth above this, the third a fourth above this, with each overtone mathematically becoming closer to the previous one.

'Why is this important to know about?' I hear you ask. . . . Well, to put it simply, as the Harmonic Overtone Series[1] continues and gets higher in pitch, some of these overtones will begin to clash with the fundamental. Dependent upon the sound source's timbre, the key of your song, and the note being played, you may wish to reduce or enhance certain aspects of a sound source to bring out or reduce certain characteristic features that you like or dislike. Understanding this concept of targeting specific elements of a sound source and why you may wish to do so will transform how you approach EQ. Rather than just fiddling until you decide you like the sound of something, you can train yourself to listen for things you like and dislike and target these specific areas.

> TASK – Load a parametric EQ onto a range of tracks within a project. Practise identifying the fundamental frequency for different sound sources.

Day 3 – Defining frequency bands

The human ear can hear frequencies between approximately 20Hz in the low end and 20,000 Hz, or 20kHz in the high end. As you get older this range narrows, particularly from the top down. So, you can expect younger people to be more responsive to the upper end of the frequency spectrum. It is for this reason that most equalisers label their upper and lower extremities as 20Hz and 20kHz. In general, you won't find many EQs allowing you to access frequencies outside of these bands.

To be able to EQ in an informed way, the best things you can learn about are the different frequency bands and the timbres and characters associated with each. The broadest way is to define sound as either low, midrange, or high end, with low being from 20 to 300Hz, midrange being 300 to 5kHz, and the high end being 5kHz to 20kHz. But this is a vast cross section and needs to be defined much further.

Your bass frequencies can be split into four bands: The deep or sub-bass is from 20 to 40Hz, low bass from 40 to 80Hz, mid-bass from 80 to 160Hz, and upper bass from 160 to 300Hz. You'll notice that each of these bands gets approximately twice as wide as the previous one. As you double the frequency, you are halving the sound wave's length and, therefore, halving the sound's energy. So, you can assume that we are essentially splitting our frequency spectrum into equally proportioned segments in terms of energy in each band that we've defined here.

The midrange bands are defined as lower midrange from 300 to 600Hz, middle midrange from 600 to 1.2kHz, upper midrange from 1.2 to 2.4kHz, and presence range from 2.4 to 5kHz. The high end is split in two, with high end from 5 to 10kHz, and extremely high end from 10 to 20kHz.

You'll often hear people describe various frequencies with characteristic adjectives such as boxy, harsh, or dull. Producers with trained ears understand that each adjective is describing a specific range in the frequency spectrum. You should grab a chart like the one shown in Table 3.1 and learn these ranges by heart. Then, when you hear a specific positive or negative characteristic in a sound, you'll know where to target to improve it.

> TASK – Make yourself a copy of my frequency chart. Copying it will do you good. Stick it somewhere prominent until you've memorised it.

Day 4 – Different types of equalisers

There are several different types of equalisers. There's the parametric EQ, the shelving, the graphic, and the dynamic EQ. And every manufacturer will tell you that theirs is the best. They'll say things like 'the most transparent', 'the most genuine emulation', 'adds beautiful warmth', 'silky smooth', and so on. But don't be fooled. All equalisers essentially do the same thing: They allow you to boost or cut frequencies. That's it! Whilst you're learning the process of EQ'ing, training your ear to recognise and identify specific frequency ranges and characteristics, there is no need to spend money on

Table 3.1 Frequency characteristics adjectively specified according to frequency point.

Hz	Low				Midrange				High End	
	20-40	40-80	80-160	160-300	300-600	600-1.2k	1.2k-2.4k	2.4k-5k	5k-10k	10k-20k
	Sub Bass	Low Bass	Mid Bass	Upper Bass	Lower Midrange	Middle Midrange	Upper Midrange	Presence Range	High End	Extremely High End
Too Much	Rumble	Boomy			Boxy	Honky/Nasty	Tinny	Harsh / Too Much Character		Brittle
			Muddy					Sibilance		
								Too Far Away		Piercing
Balanced	Bottom							Bright		
			Punch	Warm			Present		Air	
				Clear	Full		Definition	Edge		
Not Enough		Weak				Hollow			Dark	Dull
			Thin				Not Enough Character			
							Distant			

A table specifying different frequency characteristics. Across the top are labelled the frequency bands, from sub-bass across to extremely high end. Down the side are labelled too much, balanced, and not enough.

third-party plugins. Learn to EQ with the stock plugins that come with your DAW first.

With that said, let's figure out what these different types of equalisers look like. The parametric EQ is the most common sort you'll come across. It offers you a lot of control, allowing you to adjust the sound's frequency content in a wide range of different ways. Commonly, they'll have a visual representation of the soundwave within the plugin, allowing you to see how you affect the sound you are EQ'ing. Graphic EQs, in contrast to parametric EQs, have fixed frequency bands. You can only boost or cut the frequencies at the specified points on the EQ rather than adjusting the precise target frequency, as you can on a parametric EQ. A shelving EQ allows you to boost or cut frequencies in the high or low end above or below a specified target frequency. So, everything above or below the target frequency will be affected, rather than being able to home in on a specific target frequency, as with a parametric or graphic EQ. A dynamic EQ allows you to target a frequency and set a threshold for that band. When that threshold is exceeded, you can instruct the EQ to either compress or expand the targeted frequency.

By learning how each type of equaliser behaves, you will be able to select the right tool for the job. Each equaliser has its strengths and weaknesses. A parametric EQ is excellent for finding and attenuating a specific frequency but is not so good at shaping a sound's character. A graphic EQ is the opposite: Great for shaping character but not very good for targeting a specific frequency point. Shelving EQs are great for making space in your mix's high or low end but not much else. Dynamic EQs are handy for taming a specific frequency or making space in your mix at certain times without permanently losing that frequency from the sound. Learn how to use the tools you have available to you to use them most effectively. You wouldn't use a screwdriver to hammer in a nail, would you?

> TASK – Look at all the stock EQs that come with your DAW. Identify what type of equaliser they are.

Day 5 – Anatomy of an EQ 1

Whatever sort of equaliser you're using, you'll find the same anatomy across all of them. When working with a new EQ plugin, the first thing you should do is study it to understand precisely what controls you have to work with. The first control you're likely to come across is the frequency band. Sometimes these will be moveable, like on most parametric EQs, and sometimes they'll be fixed, as on graphic EQs. The frequency band allows you to select, or at least know, what frequency that band will be affecting. By knowing this and referring to our chart from before, you'll quickly assess whether a cut or boost at that frequency will provide the characteristic you're after.

The second ubiquitous control you'll come across is a filter. Lots of EQs will have high-pass and low-pass filters built in. These are self-explanatory: They allow everything above or below a specified frequency to pass through, hence the name high- or low-pass. These can be presented in several ways depending upon the EQ's style, so make sure you learn where they are. You'll use them all the time to tidy up sounds and remove undesirable frequencies in the high and low end.

> TASK – Study your DAW's stock equalisers. Identify where to adjust the frequency for each band (if applicable) and where the filters are (if it has them).

Day 6 – Anatomy of an EQ 2

Along with the filter, there are two other types of EQ you'll come across: These are the shelf and the bell. A shelf is similar to a filter in that it affects everything above or below a specified point. Unlike a filter, though, you can boost a shelf as well as cutting it, and you don't have to cut or boost all the way; as with a filter, you can just shelf a bit.

There are three other vital controls you'll come across on your equaliser. On your filters, you may have a slope control. The slope designates how aggressively the frequencies are filtered above or below your specified cut-off frequency. This is typically stated in dB/octave. Sometimes you'll want to have a steep slope to cut most of the sound, or sometimes a gentler slope may be appropriate where you wish to tidy gently and non-aggressively.

You'll also come across the Q or bandwidth control. The Q control applies to the width of a frequency band. Sometimes you'll want to target a precise frequency and will therefore require a narrow Q that only impacts on a small frequency range. Other times you may want to be gentler, most likely when boosting a frequency range, so a broader Q would be appropriate. You won't find a Q control on every EQ, though. Many EQs have bands with fixed Qs.

The other noticeable thing you'll encounter is the gain control. The gain control simply allows you to cut or boost your targeted frequency range by your specified amount. Once these controls are locked in your mind, you're 50% of the way there. As with selecting the appropriate equaliser for the job, you also want to choose the most appropriate EQ type (a filter, shelf, or bell) to do the job most efficiently.

> TASK – Explore your stock EQs further. Do any of them have shelves? If they have filters, do they have slopes? Do your bands all have Q/bandwidth control? Have you found the gain control for each band?

Day 7 – Subtractive/corrective EQ

There are only two methods of EQ'ing. The first is subtractive or corrective EQ. This applies to any cut you make. Almost all recorded sounds will have parts of their sounds that are less pleasant than others or are entirely unnecessary. Recordings pick up undesirable frequencies for all sorts of reasons, whether it's the microphone being used, the room you're recording in, or something physically within the sound source being recorded. Even synthesised sounds can have unwanted aspects to them. Identifying and reducing these undesirable or unnecessary frequencies will give you a cleaner, more pleasant sound to work with. Learning to subtract unpleasant frequencies sensitively is key to retaining the character of a sound whilst creating the space in your mix that you desire. Applying subtractive EQ too aggressively can result in removing frequencies that you would rather keep in your sound source and often means you end up sucking the life out of your sound, making it feel thin, shallow, or lacking character.

Your subtractive EQ'ing will often be the first thing you do to a sound in your signal chain. You'll want to remove unpleasant or undesirable frequencies initially before applying any other processes to a sound source. That's not to say that you can't use further subtractive EQ later in your signal chain, but it's a good idea to do it first and foremost to ensure you're working with the cleanest, best version of your sound source from the beginning.

It's important to note at this stage that any subtractive EQ move will result in a quieter sound. Reducing or cutting a frequency from a sound is literally taking that part of the sound away, so you must gain stage after applying subtractive EQ to ensure that your moves are good ones and are not just making your sound source quieter.

> TASK – Practise subtractive EQ using filters, shelves, and bells. What effect do the different EQ types have on how accurately you can correct your sound? Don't forget to gain stage accurately to match the input and output signal of the equaliser.

Day 8 – Additive/creative EQ

The second method of EQ'ing is additive or creative EQ. This applies to any frequency that you wish to boost. There will often be parts of a sound that are very pleasant that you'll want to hear more of. There's nothing wrong with boosting something that you like. However, I would suggest you proceed with this word of caution in mind: Boost with a wide Q. Keeping the Q wide when boosting will sound much more natural than using a narrow one. Narrow boosts will almost certainly sound unnatural and will end up poking through your mix in unpleasant ways. Boosting broadly means the effect is gentler and, therefore, more natural sounding.

Conversely to subtractive EQ'ing, any additive move will result in a louder signal. Therefore, you should take extra care to gain stage when doing additive EQ to ensure you level match your input and output levels. This way, you will be able properly to gauge whether your EQ moves are making your sound better and not just louder. Remember what we said back in Unit 1 about things being louder deceiving us into thinking that they're better?

> TASK – Practise additive EQ using shelves and bells. What effect do the different EQ types have on how you can shape the character of your sound? Don't forget to gain stage accurately.

Day 9 – EQ can't fix a poor recording

Before getting into specific EQ techniques, it's essential to outline a few general guidelines that will set you up for success. The first of these is this: No amount of EQ can fix a poor recording. A microphone's job is to capture the sound source as accurately as it possibly can. For this reason, microphones are highly sensitive and delicate pieces of equipment. Condenser microphones are even more so than dynamic mikes. The diaphragm within the mike will pick up the most subtle vibration to capture the sound source's precise characteristics and subtle nuances. Very often, mikes are designed to capture more than just the sound; they're designed to capture the environment in which the sound source is located.

To you, this means that before you begin to consider what you're going to do to a sound in your DAW, you must consider how to capture that sound in the best possible way to enable you to be in control of that sound at all subsequent stages in your process. If you're in the fortunate position of having more than one microphone to use, consider carefully which mike is the best one for the job. To do this, you should understand your microphones' frequency responses and characteristics and couple this with the sound you are recording. Once you've selected your mike, whether you had a choice to make or not, you must think carefully about where you place the sound source and position the microphone in relation to the sound source. Your microphone will pick up the room's sound and character and whatever it is you're recording, so record in the room or space that will give you the most control. We'll explore this a lot more in Unit 8.

Once a frequency has been burnt into a recording, you can't take it out with EQ without affecting the tone of the thing you want to hear. By spending time setting up your recording as best you can, you're making your life much easier down the line when you come to EQ.

> TASK – Read the manual(s) that came with your microphone(s). You should know about its frequency response, capsule location, maximum SPL, impedance, and polar pattern.

Day 10 – Have intent

My next broad piece of advice is this: Have an intention. I often see inexperienced producers reaching for an EQ on every single track and fiddling about with it without considering what it is they're trying to achieve. This is a good rule to adopt with every aspect of producing and mixing but is especially relevant to EQ. When listening to your mix and considering EQ, you should ask yourself, 'What do I need to do to make this sound better?' Now, the term 'better' can encompass many different things. It could be interpreted as clearer, fuller, brighter, warmer, thicker, more defined, or numerous other things. But it's by asking this question you can identify intent in your EQ'ing. For example, if you're listening to your kick in the mix and find it to be boomy, then you know you need to target the low bass frequencies. Or if you have a vocal that feels brittle, you know you're looking in the extremely high end. But without identifying intent, you will find yourself needlessly wasting time EQ'ing things that simply don't need to be touched. The common misconception is that everything needs to be EQ'ed in some way. The reality is that, whilst many aspects of a mix will benefit from some sort of EQ move, having intent will eradicate needless fiddling and time-wasting.

> TASK – Every time you reach for an EQ over the next few days, ask yourself, 'What am I trying to achieve?' Don't EQ anything without having an intention!

Day 11 – Learn the frequency spectrum

The most valuable piece of advice I can give you is to learn the frequency spectrum. The most effective way of maximising your time is to know the frequency spectrum and its specific characteristics by heart. One of the most significant differences you'll notice when watching an experienced producer versus an inexperienced one is how decisively they navigate a mix. When EQ'ing, they seem always to know precisely what they're targeting and will find the thing they're looking for with ease. This is because experienced producers understand, just by listening, what part of a sound requires attention. But it's not just about understanding whether a sound is boomy, hollow, tinny, or dull. It's also about joining the dots between these characteristics and where they lie in the frequency spectrum. By knowing by heart what frequency characteristics are likely to be problematic and where to find them in the spectrum, you will save yourself an enormous amount of time and energy.

So, print out your frequency chart, stick it on the fridge, and read it three times a day until it's committed to memory. Once you have it engrained in your mind, you'll wonder why you didn't learn it sooner.

> TASK – Make yourself a plan to memorise the frequency chart. Once it's ingrained in your mind, you'll save countless hours in the mix.

Day 12 – Use an EQ chart

Meanwhile, stick your frequency chart on the wall next to your computer, the desk, or somewhere clearly visible. It's not cheating to have helpful guides and tools around you whilst you work. These days, many producers have templates and presets, that they use frequently, saved within their DAW that they can recall at any time. Whilst you're practising a new skill, it's essential to repeatedly remind yourself of the right way to do something. The incredibly talented Julie Andrews once said, 'Amateurs practise until they get it right; professionals practise until they can't get it wrong'. Applying this to our circumstance here, we can interpret it as meaning that you should practice EQ'ing with reference to your frequency chart, EQ'ing with intent until you can do it without thought. Once you find you can fix or enhance frequencies without thinking, only once it becomes second nature can you say that it is truly embedded in your practice. And this clearly will take time and effort. You will need to invest energy into mastering this process, but the payoffs will far outweigh the invested time and energy once you have this skill nailed.

> TASK – Seriously, memorise the frequency chart. I mean it!

Day 13 – The big four: Remove the gross stuff

Earlier I said there are only two methods of EQ'ing: Additive and subtractive. This is correct. But they can be further divided into four categories. In terms of EQ, you will always be doing one of these four things. So, ensure you know which one of the four you are setting out to achieve before you start.

The first of the big four is to remove the nasty or unnecessary stuff. Generally speaking, eliminating unpleasant frequencies from your sounds will be achieved with narrow bands. A narrow band allows you to focus on a specific offending frequency and tame it without affecting everything else around it. By removing the nasty stuff from a sound, you will be making more space in your mix for other things to come through.

The first technique for doing this is known as the sweep EQ technique. What you do is to create a narrow band and boost it slightly. Then sweep it slowly across the frequency spectrum until you find a pokey frequency. I mean by pokey that, as you sweep your narrow band, some points will jump up in volume compared to other points. These are the pokey frequencies, which generally denote that too much of that frequency range is present in the sound. These pokey bits are the bits you'll want to attenuate. Go gently when you cut though. Don't feel you need to pull them out completely. Go with a 2–10dB reduction and see how you go. Some producers don't like this technique because it's easy to overdo it if you're heavy-handed. But if you proceed with caution and aim to keep things sounding natural, you'll be fine.

Removing the nasty stuff also includes high-pass filtering. You'll often find a lot of buildup of low frequencies in sounds that are entirely unnecessary and just clog up your mix. When looking on a parametric EQ, you'll be able to see the fundamental frequency and will visually see any low-end nonsense below it that isn't part of the sound you want. You can see this back in Figure 3.1. Look to remove this with a high-pass filter to keep your frequency spectrum tidy and allow your kick and bass maximum room to shine through. As a rule, gently roll up your high-pass filter until you notice it starting to affect the sound, and then back it off again until you can no longer hear it making a difference. This will ensure you're going gently enough and not removing any of the frequency content you want to keep.

> TASK – Practise high-pass filtering unnecessary low-end rumble out of some tracks. The key is to get the filter just below the fundamental so you don't lose any of the body of the sound. Be aware that the fundamental will change if different pitches are being performed.

Day 14 – The big four: Enhance the good stuff

The second of the big four is to enhance the good stuff. Now is an excellent time to introduce a guideline many producers like to follow. For subtractive EQ, many will prefer to use a transparent EQ – something that doesn't add any additional colouration or character to the signal. FabFilter's Pro-Q 3 is an industry leader, but your DAW's stock EQ will be fine. Conversely, many like to use an EQ that models an analogue EQ for additive EQ. Many traditional outboard EQs such as the Neve 1073, the API 560, and the Pultec EQP-1A are well loved for their different tones and characters they impart on the sound they are affecting. This is why producers will often select an analogue model for tone-shaping EQ moves.

In terms of the moves that you make here, you'll most likely be looking to make more of the parts of sounds that you already like. So, if you want to make more of a sound's warmth, presence, or air, now is the time to do it. Wider Qs are the way to go here. When boosting frequencies, you've already found out when you swept a narrow boost with your subtractive EQ that a narrow boost can quickly become pokey. Therefore, boosting broadly is the way to go.

Most often, this is the time to use shelves in the high end too. Rather than adding a bell and boosting at 15k to add air, for example, you may be better served to select a shelf and to roll it up gently from 10k. This will make the overall high-end boost smoother and less noticeable. For this reason, it can be excellent practice to perform tone-shaping EQ with an EQ that doesn't have a spectral analyser. This will encourage you to EQ with your ears rather than your eyes, which is a good thing. This is another benefit of using an analogue model, as these will almost certainly not have any graphic read-out of the frequency spectrum.

> TASK – If your DAW has an analogue EQ (or if you have a third-party option), explore tone-shaping with it. Replicate the EQ moves you make with a regular parametric EQ. How do they differ?

Day 15 – The big four: Make things sound different

Number three in the big four is to make things sound different. Sometimes, rather than wanting things to sound natural and enhancing the good bits, you may deliberately want to make something sound different. You may want aggressively to high-pass a sound to make it sound especially thin. You may add a further low-pass filter to this to create the well-known telephone EQ effect. You may want to use an aggressive low-pass filter to get that 'standing outside the club' feeling. Whatever it is that you're trying to achieve, there's no rulebook here. Making something sound different is about altering the natural frequency range of a sound and making it unnatural. Making things sound different from how you'd usually expect to hear them is an excellent way of adding interest to your mix and making your listener take notice.

> TASK – Experiment with filters to make things sound different. Try making the classic telephone and outside-the-club effects outlined in the previous paragraph.

Day 16 – The big four: Create space

Last in the big four is creating space in your mix. This is a critical concept, and once you've got the hang of it, you'll find your mixes slot together with much more ease. It goes like this: The human ear can only hear frequencies between 20Hz to 20kHz. This, we already know. The problem is that if you layer sounds over one another in the same frequency range, they end up masking each other, covering each other up. The result is that they just smudge into one another, and you can't hear any of them. The solution is to use the frequency-allocation principle. You specify locations within the frequency spectrum where you wish for sounds to be prominent and carve this space out of any other sounds that may conflict with this allocation. For example, your kick and bass are likely both to have frequency content present between 50 to 100Hz. Within this range, you may decide that you like the kick's sound at 60Hz and the bass at 80Hz. So, as well as creating a gentle boost at these points in the kick and bass, you would also make a small sympathetic cut in the other, effectively creating room at this frequency point for the other to shine through. You are allocating that frequency point to that particular sound. If you apply this principle to all your sounds up the frequency spectrum, you will be far more likely to arrive at a well-balanced mix, where everything has its place to be present and is sympathetic to the other elements in your mix.

> TASK – Explore frequency allocation. Try with some obvious combinations: Kick and bass, vocal and guitar, synth and keys. Can you find any less obvious pairings?

Day 17 – Don't be fooled by volume

Now let me give you some general EQ tips that should help you along the way. The first big one is not to be fooled by volume changes. With every EQ alteration you make, you will be making the original sound quieter or louder depending upon whether you are using additive or subtractive EQ moves. As we've already established, there are just four things that you may be trying to achieve with your EQ: Removing nasty stuff, enhancing pleasant stuff, making things sound different, or making space in your mix. We also know that our natural response is to perceive it as better when we hear something louder because we can hear it more clearly. But louder doesn't always mean better. As with all your mix processes, I would encourage you to gain stage after your EQ plugins, matching the input and output signal. Only by doing this can you honestly assess the success of your EQ alterations.

Almost all EQ plugins will have some sort of output gain control built in, so use this to level match by A/B'ing the plugin. If you want to be precise, you can match the exact peak level using your channel meter. There are also a few EQ plugins that have automatic gain compensation built in. FabFilter's Pro-Q 3 is one of these. But the TDR Vos SlickEQ is a great free option that automatically adjusts your signal's gain to match the input and output level.

> TASK – Practise gain staging your equalisers. Can you match your input and output signals precisely? Does this help in making better assessments of the quality of your EQ moves?

Day 18 – Small for natural, large for unnatural

Tip number two is this: If you want things to sound natural, make small EQ adjustments. If you want them to sound unnatural, make large EQ moves. The larger the EQ move you make, the more easily someone will be able to hear it in your mix. Normally, you will intend to make things sound as natural and real as you possibly can. You'll do this by gently boosting the nice stuff to enhance the natural tone and character. But occasionally, you'll want to make something sound distinctly unnatural. And this is when you may wish to utilise a narrower band.

Although slightly off-topic here, I want to draw your attention again to shelving EQs. They're much neglected among amateur producers, and they should be in your arsenal. In particular, I'd like you to explore cutting with a low shelf and boosting with a high shelf. This can often be a more natural way of shaping a sound's lower and upper parts and is likely to be less evident in your mix than a bell. Particularly, if you're gentle with the slope they can be very subtle indeed.

> TASK – Explore low cuts and high boosts with shelving EQs. Compare this with cutting and boosting at the same frequency point with a bell. Can you hear the difference?

Day 19 – Prioritise cuts

Tip three is to prioritise cuts over boosts. I will always stand by the opinion that the best way to shape a sound is to reduce or cut the less important parts to focus on the crucial components. The reasoning behind this is simple: With every boost you make to a sound, you are increasing its level. Whilst I fully endorse gain staging after each processing move, it is easy to overlook this. Therefore, especially whilst you're developing your ear and your ability quickly to identify frequencies, prioritising cuts over boosts helps to ensure that you stay in control of your mix's overall level. If you shape your mix's sound with additive EQ, your level will creep up and up, resulting in a loss of headroom, ultimately making it hard for you to get a good result out of your master.

Whilst it's impractical to suggest shaping a mix entirely with subtractive EQ, I would advise you to try and make the bulk of your tone-shaping decisions with subtractive EQ and use additive EQ to shape the character of sounds. By focusing on subtractive EQ, you are more likely to spend time creating space in your mix for things to shine through, giving you better balance overall.

> TASK – Approach EQ'ing of a mix with this new mantra in mind. Rather than saying, 'I Want more midrange in my guitar', say, 'How can I make space in my mix for the midrange in my guitar to come through?'

Day 20 – Don't EQ in solo

Don't EQ in solo. This is one of the most common mistakes I see amateur producers making all the time. People spend hours and hours EQ'ing all the individual parts of a mix one at a time only to find that they don't fit well when they put them back together again. The reason is apparent when it's explained: It doesn't matter at all what your parts sound like individually because they won't be heard separately in your mix. What matters is how it all sounds together. After all, you're going to listen to everything at the same time. You should make the majority of your EQ'ing decisions in the context of the whole mix rather than in solo. In this way, you will assess the impact of an EQ move on its relation to the rest of the mix and make an informed decision as to whether your moves are good ones.

The exception to this rule is when looking to remove specific frequencies. It can often be helpful to apply subtractive EQ moves, particularly high-pass filters, in solo. This way, you will be able to identify the frequencies you're targeting more easily. Just bear in mind that you will create space in your mix at the frequency point through every subtractive move you make. It's almost certain that something else will fill that space. Of course, this could be your goal, but on occasion, it may not be, so keep your ears open at all times and remember to check the impact of your subtractive moves in the context of the mix immediately after making them.

> TASK – Practise EQ'ing in the context of your mix rather than in solo. What impact does this have on the moves you make compared to working in solo?

Day 21 – Small changes add up

Small changes add up. I'm sure you've heard the expression, 'Save the pennies, and the pounds will take care of themselves'. In EQ terms, if you were to add a 0.5dB boost at 1k on every track, it would make very little difference individually, but in the context of your whole mix, you'd find that you end up with far too much 1k, and it would become overwhelming. The point here is to understand that EQ moves don't need to be massive to have impact. Very gently high-passing all of your tracks just below their fundamental frequency may not make much difference individually. Still, in the context of your low end across a whole track, the difference will be significant. Gently carving a little bit of 3k out of your guitars, synths, and drums may not make much difference on its own, but it may well give your lead vocal the space it needs to cut through in the presence range without requiring a significant boost in this range on the vocal. Here's a super crude way of illustrating it further: What's likely to sound more natural, making a 1dB cut in your guitars, drums, and synths at 3k, or making a 3dB boost in your lead vocal? Which one will allow you to retain more of the original sound's natural tone and character without doing anything too damaging? Once you understand and embrace this way of working, you will be well on the way to mastering EQ.

> TASK – Implement the small-changes principle into a mix. Rather than making large EQ boosts to one track, make small adjustments to others to create the space you desire. How does the result differ?

Day 22 – Subtle with stock, bold with analogue

Be subtle with stock EQ but more aggressive with analogue emulations. Without troubling you with the detail, if you use your stock parametric EQ too aggressively, it is likely to have some unpleasant side effects, such as messing with the phase of the sound. This is avoidable by using a linear phase EQ, but these tend to be very CPU-hungry. So, as a general rule, be gentle with the moves you make on your stock parametric EQ. With an analogue emulation, however, you can be far bolder. Experienced producers will select a specific analogue EQ for the colour and tone that it imparts to a sound. For example, the 'low-end trick' on the Pultec EQP-1A of boosting and cutting simultaneously at 30Hz creates a boost at 80Hz and a dip at 200Hz, which can sound great on some kick drums. The more aggressively you dial in these moves, the more of this unique character you will bring out. You can be bold with the actions you make on an analogue EQ. The circuitry they are designed to emulate imparts desirable characteristics on the tone that simply don't occur with a purely digital EQ.

> TASK – Recreate large EQ moves with a digital EQ and an analogue emulation. Can you hear the difference in tone quality between the two?

Day 23 – Don't obsess over plugin order

Very often, amateur producers will ask, 'What is the correct order to have my plugins? Should I EQ before or after compression?' The short answer is – it doesn't matter. The effect your EQ will have will differ depending on where you place it in the signal chain. EQ'ing before your compressor will have a different impact on EQ'ing afterwards. EQ'ing before your reverb send will sound different to EQ'ing later. There is no right or wrong. There is an understanding of what sound you are trying to achieve and the best way to achieve it.

I have one general exception to this rule: I would advise you to apply your main subtractive EQ moves first in your signal chain. The purpose of subtractive EQ is to remove the unpleasant or unnecessary frequencies from a sound. If you don't do this before any compression that may come later in the signal chain, you will effectively compress these frequencies into the signal and make it harder to remove them later. I always recommend making your sound as clean and tidy as possible first so that at every subsequent stage in your signal path you are only working with the desirable frequencies and none of the undesirable ones.

> TASK – Pull up a previous mix you've finished. Move your EQ before and after various plugins. Try it before/after compression, reverb, and anything else that interests you. What effect does your EQ's location have?

Day 24 – Remove the mud

A widespread issue is the build-up of unnecessary rumble in the low end. As established earlier, low frequencies carry a lot of energy. If you don't control this energy, you'll find your low end becomes a real mess, with many things fighting for attention and nothing really winning. A telltale sign of this is that your kick struggles to be heard through your mix, your bass lacks definition and just feels a bit mushy, and there's no clear definition between your bass and the other instruments in your track. You can solve this issue by high-passing things below the fundamental frequency. As a rule, the lower the sound, the more gently you want to high-pass. For example, I'll often high-pass even my bass, but just gently around 30Hz, removing frequencies that are almost imperceptible to the human ear. As instruments get higher, I'll high-pass more aggressively. So, on a lead line with a fundamental around 600Hz, I may use a slope of 18dB/octave or more to ensure it's spotless. Experiment with your slopes on your filters. The steeper the slope, the more aggressive the cutoff will become. The slope amount will also adjust the cutoff frequency.

Another prevalent issue is the build-up of the 'muddy' frequency range. We already defined this frequency range earlier, but the most common culprits will lie between 250–350Hz. To keep this section of your frequency range in check, consider applying a small cut of 2–3dB somewhere in this range in instruments that have this content present. Don't be too heavy-handed with this, or you'll end up sucking the guts out of your track, and it'll lack body.

> TASK – Find a mix that you feel lacks clarity. Try to create some clarity by cutting some of the muddy frequencies between 250–350Hz. Does this help, or is the problem elsewhere?

Day 25 – EQ in mono

I have a few more tips for you in this unit. First, try to apply your EQ moves in mono. Or at least check all your EQ moves in mono. Bear in mind that some people will ultimately listen to your music in mono through no fault of their own. So, your mix needs to translate. By checking all of your EQ moves in mono, you are forced to create space and separation in your mix. You shouldn't just rely on panning to create separation. Yes, I recommend doing your panning before EQ, as I feel this gives you a better impression of your mix and allows you to have a better balance from the outset. This shouldn't be done instead of careful frequency allocation, but in addition to it.

One of the dangers of applying EQ exclusively in stereo is that you can end up with phasing issues. For example, you may think that your guitar panned halfway left and your keyboard panned halfway right sound brilliant in stereo. And they probably do. But when summed to mono, they may conflict with each other. You need to check every element of your mix in mono truly to guarantee that your mix will translate wherever it is played.

> TASK – Practise checking your stereo EQ moves in mono. How do they translate? Pay special attention to parts you have paired off and balanced in your stereo field. These parts are most likely to have frequency build-up or phase cancellation issues.

Day 26 – Dynamic EQ

Dynamic EQ should be your best friend. Take this example: Your kick sounds excellent on its own, and so does your bass, but together they lack definition. The bass seems to be masking the kick and sucking the punch out of it. What do you do? Do you reach straight for a bell EQ and cut the frequency you like in the kick from the bass to make room for it? You certainly could do this, but then you won't have that frequency in the bass at all. And the kick isn't playing all the time as the bass is. The solution here is to use a dynamic EQ band with the sidechain input set to the kick. In this way, you can select the same frequency point where you want the kick to poke through the bass, but rather than losing it forever, you can use the kick to trigger the dynamic EQ, only pulling this frequency out when the kick strikes, and leaving it in the rest of the time.

This is a superb technique that can be used across your mix in many different ways. You could use it to sidechain duck a specific frequency in your lead guitar or synth from your lead vocal to give your vocal space to come through more when it performs. Or you could use it to sidechain duck a frequency in your vocal reverb so that it gets out of the way of the lead vocal whilst it performs but blooms when the vocal exits.

Once you get this technique down, you will not look back. I guarantee it!

> TASK – Practise setting up dynamic EQs with sidechain triggers. If your DAW doesn't have a stock EQ that can achieve this, get a third-party option. FabFilter's Pro-Q 3 is perfect, but TDR Nova is a great free option.

Day 27 – How do you pick your EQ?

With so many different EQ plugins available, and every third-party manufacturer telling you that theirs is the absolute best, how do you possibly select which one to use? There are literally hundreds available on the market. So, here's my advice on how to choose an EQ plugin.

As I've mentioned already, I would thoroughly recommend using a transparent digital parametric EQ for subtractive EQ. Your DAW's stock EQ is likely to tick this box. But if you want one with bells and whistles on it, then get FabFilter's Pro-Q 3. It's, without doubt, the industry leader for transparent EQ. For additive, tone-shaping EQ, use an analogue model of something. And this is where it gets a little grey. How do you choose which one to use? I would recommend getting one emulation of each of the following models (but get them one at a time and learn them properly. Don't let them sit in your DAW collecting virtual dust). Firstly, a Pultec EQP-1A. It's great for adding warmth and shine to kicks, guitars, and even your mix buss. Secondly, an SSL E or G channel strip. These sound great on electric guitars, vocals, and drums. Next, get a Neve 1073 or 1084. They're often described as warm, fat, and full. Try them on drums, bass, and vocals. Fourth, an API 550 and 560. They're excellent on kick, snare, bass, and guitars.

This is more than enough to get you started. You can probably find six to eight different manufacturers offering emulations of each of these models. A good option is Waves, who will often do reasonably priced bundles and discounts. I'll stress this again for emphasis: Get *one* of each and learn it. Use it lots and explore its characteristics until you really understand what tonal qualities you can get out of it. There is no right or wrong EQ to select.

> TASK – This piece of work will take some time. Get *one* of the analogue emulations mentioned previously and learn it thoroughly. Read up on it, watch videos about it, and learn its history and best applications. Be comfortable using it. Do all this before moving on to another model.

Day 28 – Unit summary

There we go then. That's Unit 3, done. And what a lot of ground we've covered! To recap, we've discussed:

- Defining frequency bands
- Different types of equalisers
- EQ anatomy
- Additive/creative EQ
- Subtractive/corrective EQ
- Having intention
- The big four: Removing the gross stuff, enhancing the good stuff, making things sound different, and creating space
- Not being fooled by volume
- Small moves for natural, large moves for unnatural
- Prioritising cuts
- Not EQ'ing in solo
- Being subtle with stock and aggressive with analogue
- Not getting hung up on plugin order
- Removing mud
- EQ'ing in mono
- Dynamic EQ

That is an awful lot of content to have covered. Don't expect all of this to have stuck in just 28 days. You should expect to revisit these notes regularly until these principles are engrained in your practice.

Coming up in the following unit, we will be exploring the topic of compression.

Checklist

- Are you comfortable with the concept of fundamentals and overtones? Can you easily identify a sound's fundamental?
- Have you learned the different frequency bands and their characteristics by heart? Do you feel confident identifying these characteristics in sounds?
- Do you know all the different types of equalisers? Do you understand their anatomies?
- Have you established a routine of intent when EQ'ing? Are you always focusing on achieving one of the big four?
- Do you gain stage accurately post-EQ?
- Are you prioritising cuts? Are you primarily EQ'ing in context? Are you checking your EQ moves in mono?
- Have you befriended dynamic EQ?
- Have you explored the different characteristics of some analogue EQ models?

You should move on to the following chapter only once you can answer yes to all these questions.

Further reading

1 Saus, W. (2022). *The harmonic series*. [online] oberton.org. Available at www.oberton.org/en/overtone-singing/harmonic-series/ [Accessed 9 Nov. 2022].

Unit 4

Compression

Day 1 – What is compression?

So, you've made it to Unit 4 in my 365-day course, helping you become a competent music producer. And we've arrived at the final process in the Big Four: Compression.

Compression is used on pretty much every single song you've ever listened to. It's something that producers reach for as second nature. But it's often misunderstood or not thought about carefully enough and therefore usually ends up not being used effectively or is implemented too aggressively.

To understand how it is poorly used or used well, we need to establish precisely what a compressor does. And the clue is in the name: A compressor squashes your sound. To be more specific, a compressor squashes the peaks of a sound, bringing them down and making them closer to the rest of the signal, therefore making the dynamic range less than it previously was.

Why is this useful? Surely to keep performances on recordings sounding natural, you want to retain as much dynamic range as possible? Well, yes, of course you want to keep your recording musical. But, when we balance multiple elements in our mix, making dynamics a little more consistent is helpful. By narrowing the dynamic range of something, you make it easier to balance your mix because the difference in level between the loudest and quietest parts of the sound is reduced.

So, what does this bad practice look like? Under-compressing means that your compressor isn't doing much work at all, resulting in a signal that is not much different from where you started. This isn't very helpful. Over-compressing means you have squashed your signal too aggressively and removed all its life and musicality. Both are bad. But as we'll find out later in this unit, choosing inappropriate attack and release times can be as damaging to your sound as over- or under-compressing.

So, let's dive in and kick things off by ensuring we clearly understand our compressors' anatomy and what all the different controls do.

Day 2 – Compressor anatomy: Ratio

There is no 'most important' control on a compressor, as it's the combination of everything working together that gets you where you want to be. But if I had to stick my neck out and pick one, I'd choose the ratio. To understand ratio is to understand the most destructive element of a compressor: I must make that clear from the outset.

The ratio defines how subtly or aggressively your compressor works. A low ratio of 2:1 is gentle, and a high ratio of 20:1 is aggressive. But what do these numbers in your ratio mean?

It's simple: At a ratio of 2:1, for every two decibels that your signal exceeds your threshold (tomorrow's topic), your signal will be reduced so that it only exceeds the threshold by one decibel. In other words, a 2:1 ratio applies compression of 50%. Two becomes one; four becomes two; eight becomes four, etc. At a ratio of 10:1, for every 10dB that your signal exceeds your threshold, your signal is reduced to 1dB. In other words, a 10:1 ratio applies compression of 90%. Ten becomes one; 20 becomes two; 30 becomes three, etc.

We can phrase this another way: A ratio of 2:1 reduces the signal by 50%, and 10:1 reduces the signal by 90%. To fill in the gaps, 3:1 is a 66% reduction, 4:1 is a 75% reduction, and 8:1 is an 87.5% reduction.

An important footnote here is that once a compressor is working with a ratio of 10:1 or more, it is considered limiting rather than compressing. This is because the amount of gain reduction is so significant that it behaves as a limiter (which cuts off all level above the threshold) rather than compressing it.

But what is this magic threshold I've mentioned a couple of times? Tomorrow, we'll take a look.

> TASK – Explore all the stock compressors that come with your DAW. Identify where the ratio control is on each. Look at some third-party options, too, if you have them. Do they all have ratio controls? Some compressors don't!

Day 3 – Compressor anatomy: Threshold

Clearly, you don't want your whole signal to be compressed. This would destroy all the dynamic range and musicality of the performance. We've established already that we want our compressor to narrow the dynamic range of a performance, not flatten it.

This is where the threshold comes in. A threshold is a crossing-over point. On your compressor, it's the point at which, once passed, the compressor will start working. Considering that we generally work with minus numbers when talking about gain staging in music-production, it is effectively off with your threshold set to zero. No part of your signal will exceed zero (unless it's clipping, deliberately or otherwise). As you reduce the threshold, the point where it reaches the highest peaks in your signal will be crossed. Once there is some signal above the threshold, the compressor will begin to work. But the compressor only acts on those parts of the signal that exceed the threshold, not on the whole signal.

It's a bit like your tax-free allowance on your income (in the UK, at least). You're allowed to earn X amount each year, upon which you don't have to pay any income tax. But once you exceed that amount, you must pay income tax on everything above that amount. But the tax-free amount is untouchable. Moving the threshold is to change the amount of tax-free income you have, or in audio terms, the amount of your signal that will avoid being compressed. The further you reduce the threshold, the more of your signal will exceed the threshold and be compressed.

To bring this full circle with the income tax metaphor, the ratio is effectively the tax rate you pay on everything above the threshold. A 2:1 ratio is a tax of 50%, meaning you give up 50% of your signal to the compressor gods. Simple, right?

> TASK – Go through all your compressors again, this time identifying the threshold controls. Do you have any models that don't have this option? Some compressors have a fixed threshold.

Day 4 – Compressor anatomy: Attack and release

Now we're on to two of the most overlooked controls on a compressor: The attack and release.

The attack is the speed at which the compressor kicks into gear once the threshold has been exceeded. The release is the speed at which the compressor returns to being off once the signal falls below the threshold. Let's look at these in turn.

The attack time denotes how punchy your compressor sounds. With fast attack times, the compressor will act on more of the initial hit or transient of a sound, reducing its impact. Take a snare drum as an example. With a speedy attack time, the drum's initial strike will be caught by the compressor and controlled, making the transient less aggressive. This transient will get through with a slower attack time before the compressor kicks in, so it won't be acted upon. Neither is right or wrong. It entirely depends on what you're trying to achieve with your compression. But it does lead us towards a general rule: If you are looking to control the initial attack of a sound because the transients are too aggressive, use a fast attack time. If you want to allow the transients through but act more on the subsequent signal of a sound after the initial strike, use a slower attack time. Faster attack times sound less natural, and slower attack times more so.

The release is the time it takes the compressor to return to its neutral position once the signal has fallen below the threshold. Understanding this statement is quite simple, but what it means in practice is less obvious. How long *do* you want your compressor to take to return to normal? How long is a piece of string? My general rule is this: On anything that is a fast, transient heavy sound (such as a drum), I look for the gain reduction meter to just return to its neutral position before the next strike happens. In effect, you are looking to make your needle bounce in time with the track. This means you allow your compressor to return to being off before the next bit of sound comes through it, so the new sound can be processed in the same way as what came before it. For less transient-heavy tracks, such as vocals or strings, I again look for a release that breathes with the part's phrasing. So, the release time will be dependent upon the track's tempo and the rhythm of the part you are compressing.

The footnote here is that lots of compressors have automatic release controls. These can be useful if you're looking to speed up your workflow and are generally reasonably reliable.

> TASK – Now, find the attack and release controls on your compressors. Do they all have them? Again, some don't.

Day 5 – Compressor anatomy: Knee and make-up gain

The last two controls to cover on a compressor are the knee and the make-up gain.

The knee is the most challenging parameter to understand. It's described as being either hard or soft. Most compressors, by default, will be set up with a hard knee. This means that the compressor will engage immediately when the threshold is exceeded. If you want the compressor to perform more gently, gradually introducing the gain reduction rather than dropping it all in, then you can go with a soft knee. My general principle is to use hard knees on rhythmic elements like drums and acoustic guitar and softer knees on more melodic instruments such as vocals or piano. You want the compression to be relatively unnoticeable and natural on melodic tracks but you are likely to want it to act fast on percussive ones.

The make-up gain, on the other hand, is very straightforward. It is simply a device for gain staging. When you compress, you reduce the peaks of your track, effectively making it quieter. Therefore, you will want to level match the input and output signal using the make-up gain. Simple!

> TASK – Identify these final elements on your compressors (if they are there). Now you can begin to explore dialling in some compression settings. Experiment with different ratios with a fixed threshold. Experiment with different attack and release times with fixed threshold and ratio. How do these parameters affect the sound?

Day 6 – The four analogue architectures: VCA

Almost all plugin compressors out there are designed to emulate one of the four analogue architectures. Each of them is constructed differently, affects the signal differently, and has different characteristics. I'm going to take you through them all now.

The most ubiquitous is the Voltage Controlled Amplifier – the VCA. The VCA is known for not colouring the sound that passes through it and is loved for beefing up transient-heavy content such as drums. This is because it responds extremely fast and is incredibly transparent-sounding.

In a VCA compressor, the signal path is split in two. The first path, the detector path, controls the compression effect. The second path, the output path, is what you hear. The signal path can be manipulated precisely, making it a highly accurate model.

VCA compressors are found on SSL channel strips, API compressors, and some products from Neve. They're known for being predictable and repeatable and are often employed on the master buss and group channels.[1]

> TASK – Do some further reading on VCA compressors. What are the most famous models of VCA compressors? What are they most often used for?

Day 7 – The four analogue architectures: Optical

The second analogue architecture to look at is the optical compressor. Optical compressors are interesting because they rely on light-dependent resistors. The audio signal coming into the compressor feeds a light source, usually an LED. As the input signal varies, so does the brightness of the LED and, in turn, the amount of resistance generated by the light-dependent resistor. The beauty of optical compressors is that, whilst they react quickly, they're not immediate.

The most desirable characteristic of an optical compressor is that it works in a non-linear fashion. Non-linear means that the way the attack and release operate is not a straight line with a consistent pace but rather it is curved, with a variable rate. For example, there's often a slight delay before the attack kicks in. The harder you hit an optical compressor, the faster its initial release time will be. The return to being off will be sloped, with the release getting slower and slower as it falls.

The result of all this is that optical compressors are very smooth and therefore musical. They don't jump or jolt but rather they glide. Consequently, they're great on vocals and other melodic elements that require 'rounding out'. The classic optical models are the LA-2A and 3A. Note how they lack attack and release controls.[2]

> TASK – Do some further reading on optical compressors. What controls are obviously missing from a 2A and 3A? What are they most often used for?

Day 8 – The four analogue architectures: FET

FET stands for Field Effect Transistor. A transistor can both attenuate and amplify sound depending upon the settings you dial in.

People often wonder about the differences between FET and VCA-style compressors. The difference lies in the transistors: In a VCA compressor, the transistor responds only to the incoming signal's voltage. In a FET compressor, the transistor responds to electrical charges and voltage.

As a producer, you don't need to get too hung up on this. What is essential to know is that one of the most famous compressors of all time, the 1176, is a FET compressor. Its reputation is for rapid attack times with desirable colouration. Typically, FET compressors don't have a threshold control. They work by driving the input and balancing the output to achieve your desired result.

Therefore, they're fantastic to use on guitars and drums, both individual channels and busses, but not so suitable for your mix buss.[3]

> TASK – Explore the 1176 compressor. What controls are obviously missing from a 1176? What are some of the main differences between the different revisions of the compressor?

Day 9 – The four analogue architectures: Variable Mu

Variable Mu compressors rely on valves (or tubes if you're in the USA). As the signal entering increases in these compressors, the current being sent to the valves decreases, resulting in the overall reduction in level. The valve circuitry in these compressors (and the pleasant distortion it imparts) and their highly responsive nature earn the Variable Mu the reputation for being thick and creamy.

Variable Mu compressors can handle large amounts of gain reduction before any noticeably obvious nastiness is introduced to your signal. Therefore, they are best implemented on your mix buss to 'glue' everything together. They handle transient-heavy material smoothly rather than aggressively (like the VCA and FET models).

Your poster boy models here are the Fairchild and the Manley. Notice how most Variable Mu-style compressors lack a ratio knob.[4]

> TASK – Explore Variable Mu compressors. Listen to some examples on YouTube. What other applications might they be good for?

Day 10 – Use one compressor to learn

Now that we've established what everything on a compressor does and what the different types of compressors we'll come across are, it's time to learn how to use them properly.

Having just bombarded you with a whole heap of somewhat technical information, I advise you not to worry about any of it. At least initially. Whilst you're learning to compress correctly, discovering how to dial in attack and release settings to achieve your desired sound, training your ear to level match accurately, etc., stick to one compressor. Whatever stock compressor came with your DAW will be fine for now. Until you've learned to hear how a compressor is behaving and its effect on your music, you won't benefit from having a range of different compressors. Only by training your ear to hear the subtle nuances a slight adjustment can make will you be able to implement the different architectures effectively.

So, pick one compressor and use it on everything. Only when you feel you've nailed the process of dialling in your desired compression settings every time should you consider filling your toolbox with lovely emulations. And, trust me, you'll appreciate them so much more when you've invested the time learning to hear what impact they have.

> TASK – Practise compressing elements of your mix using one compressor. Remember, compression is about controlling dynamics. If you feel you're changing the overall tone of something, you're probably doing too much.

Day 11 – What comes first: Compression or EQ?

This is music-production's version of the chicken-and-egg scenario. It's a question that doesn't have a definitive answer. In other words, you can argue it either way, and you'll never get to a conclusion because there isn't one!

Compression can be applied before or after EQ depending upon the sound source you're working with and what you're trying to achieve. I mentioned in the previous unit that I advise you to do any subtractive EQ'ing (as required) before compression, but other than that, the jury is out.

Let's consider some scenarios: You're working with a drum performance, and, for whatever reason, the dynamics are all over the place. It's sensible to narrow this dynamic range with compression before EQ'ing. What if you're producing a vocalist with a particularly prominent frequency in their voice that is triggering your compressor excessively? EQ first.

The beauty of working in the digital realm is that you can apply as many stages of EQ and compression as you require. So, if your post-compression EQ is adding new peaks that need further taming, compress again. If your post-EQ compression brings out more unpleasant frequencies that you hadn't noticed before, EQ again!

> TASK – Experiment with your EQ and compression order. Practise it both ways. What differences does it make to your workflow? What differences does it make to your sound?

Day 12 – Don't compress just because you can

Compression was first invented as an alternative to riding the volume fader. Imagine having to sit there and manually ride every single fader in your project, responding to every part of the sound that was too loud or not loud enough. That's the definition of a nightmare right there!

Being able to turn to a compressor to do this job for you is a game-changer. But do you need it on every single track in your project? To best consider this question, you should refer to the scenario where you're riding every single fader. Would you seriously sit there and ride level changes on every single channel in your project? No, of course you wouldn't. It would take forever. You would carefully select the key elements that genuinely need a more consistent dynamic and ride those.

You should approach compression in the same way. Just because you *can* throw multiple compressors on every channel in your project doesn't mean you should. Dynamic variety is good. It's what gives your music life, humanism, and breath. To flatten the dynamic range of everything is to suck the life out of it.

Asking, 'How much do I need to compress this?' is good. But first, asking, 'Do I *need* to compress this at all?' is even better.

> TASK – Revisit a previous mix. Bypass all your compressors. Analyse each channel by asking yourself, 'Do I need to compress this?' If you can't hear parts of it because it drops too low, the answer is yes. Otherwise, leave it alone!

Day 13 – Parallel compression is the best of both

Before establishing why parallel compression is a good thing, I must clarify what it means to do something in parallel.

When you place a plugin as an insert on your channel strip, you process your sound in series. Like a TV series (or season) where you chronologically watch one episode after the other, your signal passes from one plugin to the next. In contrast, parallel processing means doing something simultaneously. Imagine watching episodes one and two at the same time. They're intrinsically linked to each other by the same plot but are pretty different.

What are the benefits of compressing in parallel rather than in series? Well, if your original performance is expressive with quite a wide dynamic range, it's unlikely that you want to destroy this by compressing too hard. But if it's getting lost in the mix at specific points, you will want to compress it somehow. By compressing in parallel rather than in series, you retain all the original dynamic variety whilst having a new, flatter signal to blend underneath the original to support it. Vocals are a great candidate for this.

Let's take drums as another example. Drums are transient-rich instruments. The problem is that your transients will be much louder than the main tone of the drums. When compressing in series, reducing the differential between the transient and the sound's main body can be challenging without destroying the transient attack. Instead, compress in parallel. This way, you can retain the original transient attack but blend in a compressed signal that focuses on bringing out the drums' body rather than the front end.

It's worth noting that many compressors have some sort of blend or mix knob, allowing you to dial in your compression in parallel within the plugin itself.

> TASK – Explore parallel compression in a mix. Start with it on busses. How does it help to thicken a sound? Then explore its application on individual components. Try kick, snare, and lead vocal.

Day 14 – Matching input and output levels accurately

We've covered gain staging multiple times in this course, so I won't bore you with reminding you of its importance again at this point. However, gain staging within a compressor is a little bit different.

The reason for this is simple: Compressors are designed to reduce the dynamic range of your sound. This is different to how any other plugin works. An equaliser, for example, is straightforward: If you have added 3dB to your peak level whilst EQ'ing, you need only remove 3dB from your output to match the level.

In a compressor, because it is altering the whole dynamic performance, you will find that your compressed level is louder if you gain stage to the same peak level post-compression. This is logical when we consider that a compressor reduces dynamic range. If you match the input and output's peak level, then, of course, because the whole dynamic range has been narrowed, it will now be louder across the board.

So how do you gain stage post-compression? The thing you need to consider here is relative perceived loudness. And how do you measure that? Remember that VU Meter that you set up earlier? There you go! Where the needle peaks on heavy transients is irrelevant. You should look for the needle to be roughly hovering in the same position during the sound's main body. You should be able to A/B your compressor and not notice any perceivable loudness differential. The difference should be entirely in the consistency of level and the tone and shape of the sound. Ultimately, I'm talking about trusting your ears here.

Only by gain staging the *right* way will you legitimately be able to gauge whether your compressor is achieving what you desire.

TASK – Practise gain staging your compressors using a VU meter. The more you practise, the less you'll need to rely on the meter, and the more you can rely on your ears.

Day 15 – Avoid presets and solo

Presets are helpful for many things, and, in general, I'm a fan of them. But not when it comes to compression. Let me explain this: Plugins that don't affect dynamics are prime candidates for presets. If you want to store that 50Hz high-pass with a 500Hz scoop for your kick drum, no problem. If you're going to store that quirky dotted 1/8th note ping-pong delay, no problem. Because what they're doing to the sound has no bearing on input or output signal, you can recall these presets without issue and tweak them accordingly. The problem with dynamic processing is that how the compressor responds depends on the signal coming into it and many other factors.

The first consideration is the input level. For a preset to be of value, your input signal would need to be identical for the preset you've stored to work in the same way as it did previously. Sure, you can just go back to the input level and adjust or alter the threshold – no big deal. But then you need to consider the attack and release time. Unless the actual shape of the sound coming in is also identical, your attack and release settings will need to be tweaked too. And by the time you've done all that, you might as well have started from scratch. That's my view, anyway.

As for applying compression in solo, here's my thinking: Compression is used to assist in allowing an element to sit better in the context of your mix. If you were going to listen to something in isolation, you would not need to compress it as it would have nothing else to compete with. Therefore, the compression only matters in the context of the mix, not out of it. The question you should ask yourself is, 'Am I getting the consistency in dynamics required to make this sit in my mix better?' You can only answer this question if you are listening in the context of your mix. With that being said, I understand that to hear the subtle nuances of the attack and release, it is beneficial to check in solo. But try to train yourself to make the final decision on your compressor settings in the context of the mix and not in solo.

> TASK – Practise applying compression in the context of your mix. Sure, you'll want to check it in solo, but remember, you won't be listening to it in solo at any other point. So, try to keep it to a minimum.

Day 16 – Stacking compressors

Stacking compressors means using multiple compressors in series. And there's a good reason as to why you may wish to do this.

Let's say you're working with a vocal that has an extensive dynamic range: Say, a differential of 15–20dB between the loudest and quietest sections. You will certainly want to narrow this gap to make it manageable within your mix. But when you apply the 10–12dB of gain reduction you're looking for, you find that it sounds unnatural. Of course it does. You're reducing the dynamic range by 12dB in one hit!

A better approach would be to use two compressors in series, each aiming for 3–6dB of gain reduction. In this way, you gradually narrow the dynamic range of your track rather than slamming it. Thus, it's far less likely to sound unnatural.

But there's a bonus to compressing in series, which is that you can use different compressor models. Sticking with the vocal as our example, consider placing a FET compressor such as a 1176 first in the chain. Set it to work on the most prominent peaks in your signal, grabbing the jumpy bits quickly to bring them back in line with the rest of your signal. Then follow this with an optical compressor such as an LA-2A. With this compressor, you can gently round out the sound in a musical way. You can benefit from the familiar characteristics of different compressor architectures in this way whilst gradually achieving a more desirable result.

> TASK – Practise serial compression on vocals. If you have a 1176 and LA-2A emulation, great. But you can still implement the technique with two stock compressors in series.

Day 17 – Compressing low end

The low end in any track is an important thing to get right. Your goal should be to make it solid and consistent so that there are no significant fluctuations in level. If you have moments where the bass is suddenly much quieter, it will feel like the guts have dropped out of your track; it will feel spineless and weak.

A common technique to assist with this is to compress the kick and bass elements of your arrangement together. In this way, you can gently squeeze them, helping glue them together, making them sit together more comfortably. You don't want to be too aggressive initially. Look for just 2–3dB of gain reduction on your compressor.

If this isn't doing it for you though, you can try being more targeted. Instead, or in addition to the previous compressor, add a multiband compressor to your bass buss. Look to compress everything below around 80Hz but leave the rest of the signal untouched. This will allow your kick and bass's attack and energy through, retaining their dynamic range, but the low frequencies will be squeezed together. In this case, look for circa 5dB of gain reduction.

> TASK – Experiment with this technique in different genres. It is more appropriate in some circumstances than others. Where do you find it working most effectively?

Day 18 – The extreme-threshold technique

Dialling in the appropriate attack and release times is probably the most challenging aspect of compression. The most minor adjustments can make or break the shape of your sound. Having your attack too slow may mean you don't get the punch you want, but too fast, and you crush your transient. It's a delicate balancing act.

To help with this, you can employ the extreme-threshold technique. It's straightforward: Just reduce your threshold way too far. This will allow much more of your signal to exceed the threshold and trigger the compressor, giving you more signal to hear the impact that your attack and release settings are having. You can dial in your attack and release times from here. Then, once you're happy, back your threshold to a reasonable level where you achieve the gain reduction you desire.

This technique is advantageous when trying to learn a new compressor's characteristics.

> TASK – Practise the extreme-threshold technique. Try it on different sound sources. Is it more useful on drums, bass, vocals, or something else?

Day 19 – Don't kill the transients

Be cautious when dialling in your attack time. If you make it too short, you will cut off the transient. The effect this has is to kill the attack and punch of the sound. Sometimes, this may be your desired outcome. But more often than not, it won't be. Especially on drums, you should proceed with caution. Every sound requires a slightly different attack time, but if you're worried, then err on the side of caution. Better the attack is too long than too short.

With that said, a reliable method is to start with a slow attack and then gradually reduce it until you can noticeably hear the transient being cut off. Back it off from there until you have the transient back, and you'll be in a good place.

> TASK – Spend some time practising getting your attack time right on drums. Focus on ensuring you retain the punch and impact of the drum.

Day 20 – A little here, a little there

Music-production is about doing the little things well. I talked in the previous unit about the accumulative impact that small EQ moves have in creating space in your mix. Compression is no different.

The more control you have over your signal at multiple stages in the signal chain, the more balanced your mix will be when the signal reaches the stereo out. Let's consider a lead vocal as an example. Say that you've applied compression in series on your vocal track, first with a FET and then with an optical-style compressor. Your FET is doing 3–5dB of gain reduction, and so is your optical. This gives you 6–10dB of gain reduction overall. But then you have a double track and a triple track that enter in the chorus. So, you buss these three parts and gently compress them together to glue them. You use a VCA here, aiming for 2–3dB of gain reduction at most. But then you have a load of harmonies too. These come in and out at various points in the track. So, you create a vocal buss and send both your lead vocal buss and your harmonies buss to this. Again, you compress gently once more, this time with a Variable Mu. Here you're looking to join them all up, warming them nicely with the valve emulation and making them feel like one performance rather than multiple individual tracks. You only want 1–2dB of gain reduction here, but the additional valve emulation in the Variable Mu provides the extra harmonic richness that beautifully ties all the vocals together.

The objective with all these small moves is to gradually work towards an end goal that is well controlled. 2dB here and 1dB there won't have much of an impact on its own, but accumulatively it all contributes positively.

> TASK – Work on adding compression subtly at multiple stages of a mix rather than applying it aggressively in one go. Consider the different analogue architectures. If you have different emulations, think about what is best suited at each stage of the signal chain.

Day 21 – The right compressor for the job

With so many options available to you, it can be tricky to navigate plugin selection complexities. As I mentioned earlier in the course, the market is flooded with plugin manufacturers telling you that their product is the answer to all your problems. But don't be fooled into blowing your hard-earned cash without careful consideration.

To me, the art of music-production is about selecting the right tool for the job. And to do this well, you must fully understand the tools you have at your disposal. Yes, the more tools you have in your toolbox, the more likely you will have the right tool for the job, but you don't really need more than one hammer, do you?

In terms of compressors, once you fully understand how they work and how to dial them in to achieve what you desire, I would recommend getting one emulation of each of the four architectures. So, one VCA, one FET, one optical, and one Variable Mu. This will undoubtedly be enough to get you where you want to go, at least whilst you properly learn their individual characteristics.

But having the tools doesn't help you to select the right one. To do that, you need a quick way of remembering what they all do. So, here's a snapshot:

- VCA – great for uncoloured control, beefing up transient-heavy content, fast response, very transparent.
- Optical – great for melodic parts, vocals, synths, strings, etc., as they're highly 'musical'.

- FET – super-fast, impart desirable colour, great for guitars, drums, and aggressive vocals.
- Variable Mu – valve sound is rich, performs slowly and is super smooth, great for groups and master busses.

> TASK – Write out these characteristics and learn them. By knowing what each architecture does best, you'll be able to select the right tool for the job more quickly.

Day 22 – Don't be fooled by volume

I mentioned this earlier in the unit, but it's such an important point that I want to labour it here. You know already how your brain perceives louder things as better. And when compressing something, it's straightforward to increase the make-up gain too much, making your compressed signal more audible than the input signal and therefore fooling yourself into thinking you've made things better.

But with compression, the ramifications of being fooled by volume differentials are perhaps more significant than with other processes in your signal chain. Why? Because when you hear something sitting more consistently and prominently in your mix, your natural response is to assume that you've hit the nail on the head.

In reality, compressors do much more than just even out dynamics. How they shape your sound's tone with the attack and release controls is as important as how much gain reduction they are doing. With emulations that impart harmonic distortion and colouration to your sound, driving the right amount of signal through the compressor to achieve a good amount of saturation without going too far is a delicate balancing act.

To hear the contribution these features are having to your sound, you *must* make sure you have gain staged correctly, matching your input and output levels. If you've forgotten how I advised you to do this, flick back to Day 14 and reread!

> TASK – Spend some more time focusing on your gain staging. It's such an essential skill that it's worth spending the extra time.

Day 23 – Complementing compression with transient enhancement

Getting the attack setting just right on a compressor is challenging, even for more experienced producers. As you build your mix, gradually introducing more elements into it, something that sounded great before may not sound so good later. As you know from our discussions on frequency allocation, as frequency areas build up and become congested with different sounds, it becomes more difficult for things to cut through the mix and be heard clearly.

In particular, you will often find that, as your mix grows, the front end or attack of sounds becomes less defined, becoming a little blurred.

You may think that reaching back for the attack dial on your compressor is the way to go, and it could be. But you have probably already invested some time in dialling that in to shape the sound in a way that is pleasing to you. So perhaps adjusting that again now isn't the way forward.

Instead, you can reach for a transient shaper. Transient shapers are a secret weapon for producers. Their job is to shape the transient response of a sound without affecting the overall level. Unlike compressors, transient shapers transparently sculpt the shape of the sound from attack to sustain. They traditionally just have two controls on them: Attack and sustain. So, if you find that you've suddenly lost the punch of your kick or the thwack of your snare, grab a transient shaper and try dialling in the attack to help it poke back through your mix.

> TASK – Get yourself a transient shaper and experiment with it on transient-heavy material. Think drums, acoustic, rhythm guitar parts, etc. There are lots of good options on the market.

Day 24 – Sidechain compression

Sidechain compression is one of the most popular techniques used by the modern producer. Many find it challenging to grasp because it's often explained poorly, but it's simple.

Until now, we have talked about compressors acting upon the signal of the sound you want to compress. For example, the compressor on your bass track is listening to the bass track itself and responding to that signal directly.

But this isn't the only way a compressor can behave. Instead of responding to the sound you want to compress, you can instruct it to listen to a different sound and compress your signal based upon what the other sound is doing. This is called sidechaining.

The most common example of this is setting your bass compressor's sidechain input to your kick drum. The compressor then listens to the kick drum rather than the bass. You dial in the compression amount you want using the kick as the input signal, meaning that the compressor will act when the kick exceeds the threshold, not when the bass does. But the compressor will still act upon the bass, not the kick. This allows you to achieve a ducking effect, where the bass ducks out of the way of the kick drum when it strikes.

Therefore, the primary function of sidechain compression is to create space in your mix for things to poke through. It's the compression equivalent of frequency allocation for EQ.

But what are some other uses for sidechain compression beyond the obvious kick/snare-ducking cliché? You could use it to sidechain your lead vocal to your lead synth or guitar. In this way, you can use the lead vocal to duck the synth sound out of the way whilst it is present to ensure it has room to come through the mix. Similarly, you could use your lead vocal to sidechain your reverb or delay effects, again ducking them out of the way of the vocal whilst it's present but allowing them to bloom after the vocal exits. Or what about sidechaining your riff to your pads? Your riff doesn't play all the time, but when it does, you want the pad to get out of the way.

There are loads of creative uses for sidechain compression, and I haven't even mentioned using a multiband compressor to apply the sidechain compression to just a specific frequency range!

> TASK – Practise setting up sidechain compression. Experiment with it in a variety of different applications. Begin using it too aggressively so you can really hear the effect, and then dial it back to a more tasteful level.

Day 25 – Sidechain EQ

The world of sidechaining doesn't end there, though. Some compressors will also come with a sidechain EQ function. In the same way that sidechain compression allows you to use a specific (external) source to trigger the compressor, sidechain EQ enables you to use a particular frequency or frequency range to affect how the compressor behaves. There are two main types of sidechain EQ.

Firstly, you may come across a high-pass filter on the sidechain. This filter allows you to roll off the bottom end of a signal, meaning that the low frequencies will no longer trigger the compressor. Note that this high-pass filter isn't filtering the sound you're hearing, only the sound the compressor is responding to. This technique is often used when compressing busses or whole mixes to prevent undesirable pumping produced by the kick and bass triggering the compressor.

Secondly, you may come across a dedicated sidechain EQ section in your compressor. This will allow you to boost or attenuate specific frequencies feeding your compressor to make it respond more to particular frequency areas. This technique is used often to de-ess vocals. By raising the sibilant frequencies in the compressor's sidechain EQ, the compressor will then act more on these sibilant frequencies. Another example is on drum room mikes.

You can boost the upper frequencies in the sidechain EQ, so the compressor works more on the cymbals, but reduce the low frequencies being fed in, so the shells are not acted upon as much.

As you can see, sidechain EQ is a powerful device for shaping how your compressor responds to your sound. Don't expect to nail this technique overnight. It's advanced stuff!

> TASK – Experiment with sidechain EQ. Your DAW's stock EQ may well have this feature built in. Explore some extreme settings so you can clearly hear its effect on your compression.

Day 26 – Colouring your sound

In using an analogue emulation of a piece of hardware, many plugin compressors impart harmonic distortion that is really pleasing in a mix. But what is this colouration and harmonic distortion that is so desirable?

Consider this metaphor: You go to the cinema to see a groundbreaking new movie. You come out exactly the same person you were when you went in, at least physically. But mentally, you are changed. The characters and the plot have had a profound effect on you and have altered you inside forever. This is what happens when a signal passes through any analogue hardware. It goes in and comes out looking, to the untrained eye, the same. But look closely and you'll see that the electronic components within the compressor have subtly enhanced and enriched the sound's character.

Harmonic distortion, colouration, or saturation, as it's commonly called, is the addition of harmonics above the fundamental. Which harmonics are added, and how many, is dependent on the construction of the hardware itself, or in our case, the plugin emulation. The good thing about working in the digital domain is that sometimes the plugins offer you control over how much of this colouration is imparted onto the signal, giving you complete control over the saturation level.

And why does it sound so much better in your mix? Well, because with the addition of extra harmonic content, there is simply more to the sound. There are more frequencies present, allowing it to sound fuller and more present across more of your mix.[5]

> TASK – A/B the same compression settings with a stock digital compressor and with an analogue emulation. Can you hear the difference between the two? Does one sound better to your ear?

Day 27 – Multiband compression

The last component we'll look at in this unit is multiband compression. Now, many people find this topic to be completely baffling, but it needn't be. Multiband compression is just a more targeted approach to broadband compression, which is the full name for the sort of compression we've been discussing until now.

Multiband compression allows you to set a specific frequency range for your compressor to operate in and will leave everything else untouched or operating under its own settings. Consider some scenarios where this may be useful.

You have a singer with a cracking voice, but when they really go for it in their upper register, there's a frequency that jumps out. Multiband compression is perfect for isolating this and taming it. You're mastering your track and notice that the bass jumps out at a couple of different points. Multiband compression can solve this. You're working on an acoustic guitar track that, for the most part, sounds lush. But it has a peculiar resonance when it plays a specific chord. Multiband compression will fix that too.

The only additional stage in your process here is to identify the frequency range that you want to target and then treat it as you would any other compressor. Your multiband compressor should come with the ability to solo specific bands you create, so you focus on isolating the exact range you want without compressing additional content.

It will take some time to get used to using a multiband compressor, but don't let that deter you. Proceed cautiously; go with a gentle touch, and you won't ruin anything. As with all new skills, practice makes perfect.

> TASK – Explore some contrasting implementations for multiband compression. Perhaps on your low end, maybe on your muddy frequencies, possibly on taming a specific resonance.

Day 28 – Unit summary

Congratulations! You've reached the end of the Big Four. The great news is that once you've nailed these four fundamental skills, you're 90% of the way to stand-up mixes every time. As a quick recap, in this unit we've covered:

- The anatomy of a compressor
- The four analogue architectures
- Gain staging correctly
- Series compression
- Parallel compression
- Avoiding presets

- Compressing low end together
- Looking after the transients and transient shapers
- Compressor selection
- Sidechain compression and EQ
- Multiband compression

This unit has been especially technical. I expect you'll want to go back and reread some chapters to consolidate what you've learned.

But what's next? The last 10%, obviously! And this is where the exciting stuff happens, where you get to be creative rather than technical. Next up, we'll look at creating depth in your mix with time-based effects.

Checklist

- Are you comfortable with what compression does and the anatomy of a compressor?
- Are you familiar with the four analogue architectures? Can you select the most appropriate one for the job?
- Are you confident in identifying where best to compress in your signal chain?
- Have you explored parallel compression and understood its benefits?
- Have you practised gain staging post-compression extensively?
- Do you understand the benefit of serial compression?
- Have you experimented with transient design to complement your compression?
- Have you conquered the challenging subjects of sidechain compression and sidechain EQ?
- Have you begun to explore the benefits of multiband compression vs broadband compression?

You should move on to the following chapter only once you can answer yes to all these questions.

Further reading

1 Allen, E. (2016). *Vintage King's guide to VCA compressors*. [online] vintageking.com. Available at https://vintageking.com/blog/2016/03/vca-compressors/ [Accessed 9 Nov. 2022].
2 Sibthorpe, M. (2022). *Optical compressors explained*. [online] fortemastering.com. Available at https://fortemastering.com/optical-compressors-explained/ [Accessed 9 Nov. 2022].
3 Fuston, L. (2022). *UA's classic 1176 compressor: A history*. [online] uaudio.com. Available at www.uaudio.com/blog/analog-obsession-1176-history/ [Accessed 9 Nov. 2022].
4 Fox, A. (2022). *What is a Variable-Mu (tube) compressor and how does it work?* [online] mynewmicrophone.com. Available at https://mynewmicrophone.com/

what-is-a-variable-mu-tube-compressor-how-does-it-work/#:~:text=What%20is%20 a%20variable%2Dmu%20compressor%3F,reduction%20in%20the%20overall%20level [Accessed 9 Nov. 2022].
5 Haroosh, I. (2021). *What is saturation? And how to use it* [online] wealthysound. com. Available at https://wealthysound.com/posts/what-is-saturation-how-to-use-it [Accessed 9 Nov. 2022].

Unit 5

Reverb

Day 1 – What is reverb?

Congratulations on making it through the Big Four. Now that you've got the fundamentals of mixing under your belt, it's time to get creative. And what better place to start than with the most used effect in the audio world: Reverb.

But what is reverb? Reverb is short for reverberation, which is the act of sound waves reflecting off surfaces. Reverb is around you all the time. How noticeable it is will depend on several environmental factors. These include how hard or soft the reflective surface is, how big or small the space you're in is, how empty or full that space is, how much background noise there is competing with the reverb, and how loud the original sound is to begin with. You will hear it very noticeably if, for example, you go into a cathedral, which is a vast, open space, most likely made of stone, which is a shiny, hard, highly reflective material. A sports hall or gymnasium would display similar characteristics. Being outside in a field would be the complete opposite, where the nearest reflective surfaces could be hundreds of metres away. Alternatively, being inside a small vocal booth with tight, soft, padded walls and a low ceiling would provide another starkly contrasting reverb.

Imagine you are in a school hall (see Figure 5.1). You are standing in the middle of the hall with a drummer on the stage. The drummer bangs the drum, and the sound travels directly from the drum into your ears. This is the direct sound. But that same sound doesn't just travel directly from the drum to your ears. It is emitted in every direction in equal measure, just like the ripples created when you throw a stone into still water. As these ripples, or soundwaves, hit the hall's walls, floor, and ceiling, they are reflected. These reflections arrive in your ears at different times after the direct sound. They will always come after the direct sound as they've had further to travel. This is reverb.

Day 2 – Why do you need reverb?

Every sound you hear is subconsciously informed by the space in which it is heard. Your brain is so used to hearing sound within the context of a space

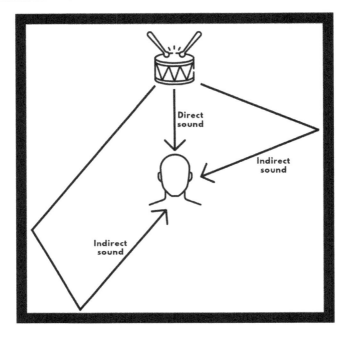

Figure 5.1 Illustrates the differences in time between direct and indirect sound.
Source: Image by Sam George.

that things sound unnatural, hollow, and thin without it. So, the simple answer to the question is that you need reverb for something to sound natural.

So why do recording studios work so hard to treat their spaces to remove as much reverb as possible? If our brains desire reverb, indeed, we should just leave all our recording spaces untreated. This would be a *much* cheaper solution!

In some cases, this is precisely what happens. As a child, I was lucky enough to be taken to the amphitheatre at Epidaurus. My Dad told me to stand right at the top of the theatre in the seats furthest away from the playing area, and he stood down in the centre. He then spoke to me at barely more than a whisper. And to my surprise, I could hear every word he said with perfect clarity. It turns out the Ancient Greeks understood acoustics pretty well, as they deliberately built their theatres to capture as much sound as possible to allow every patron to hear the players, however cheap their seats were.[1]

But of course, the Ancient Greeks had no concept of recording this sound. Their goal was simply to ensure everyone could hear what was going on. The issue with untreated acoustic spaces is that they will only ever sound like themselves. If you're AIR Studios in London or the Royal Albert Hall, this isn't such a bad thing. Having the characteristic of those spaces baked into your recording is highly desirable. But the issue is that once the natural reverberation of a recording venue is committed 'to tape', it cannot be removed. It's there forever. So, it will only ever sound like that space. The reason people desire clean recordings with no noticeable reverb is so that they can add reverb later. In this way, they can add whatever reverb they want and make it sound like it was recorded in the Sistine Chapel, the Sydney Opera House, or Ronnie Scott's, if they so wish.

> TASK – Consider some spaces or locations that you know well. It could be your kitchen, workplace, or a venue you play often. What do these spaces sound like? Write down their characteristics. How do they compare?

Day 3 – What does reverb do?

What does reverb do in a musical context? How it works is a whole complex heap of mathematics. If you're interested in this, read up on Sabine's Equation.[2] To pre-warn you, Sabine's Equation is displayed in Figure 5.2. It's not for the fainthearted! But I'm not here to bamboozle you. So, let's simplify it as much as possible. How you perceive reverb is a combination of four things: The reverberation time (RT), the volume of the room (V), the speed of sound (C), and the absorption of the room (S). And you'll find most of these controls on any reverb plugin. The reverberation time may be called length, the volume of the room may be called size, and the absorption of the space may be displayed as a choice of materials from which the reflective surfaces are made, but they mean the same thing. However, the speed of sound won't be in your plugin, as it's assumed that this is a constant.

Reverb provides depth and space to your music, but it also includes information as to where the sound is taking place and where the listener is positioned in relation to the sound. Therefore, reverb is the vehicle that transports your listener to a concert hall, a studio, a bathroom, or anywhere else. It's vital in providing realism and making your mix sound convincing.

$$RT_{60} = \frac{24 (\ln 10) V}{C_{20} S_a}$$

Figure 5.2 Image shows Sabine's Equation.
Source: Image by Sam George.

> TASK – Examine the reverb plugins you have. Can you identify the reverb parameters mentioned previously (time, volume, absorption)?

Day 4 – Different types of reverb: Hall

There are many different types of reverb. A problem many young producers struggle with is what sort of reverb to select. They struggle with this decision because they haven't learned about the characteristics of each reverb type. Over the next few days, I will take you through each reverb type individually to ensure you don't make this mistake.

We'll start by looking at hall reverb. Hall reverb is one of the most used reverb types in modern music and sound production. It simulates the acoustics of a concert hall or another similarly large, reverberant space. Halls are generally designed to produce a highly desirable listening experience. If you have ever been lucky enough to listen to music in the Royal Albert Hall, for example, you will have experienced a sound that was tonally very even, not enhancing or attenuating any particular frequency point too much. Sometimes, halls are designed to boost low frequencies. This is to make large orchestras sound even more exciting. Concert halls generally have bow-shaped architectures, which helps with their unique sound. Because of their size, they tend to have long delays, sometimes up to a few seconds long. But with this length comes thickness and layering, which can sometimes muddy your mix if not used sparingly.

For these reasons, hall reverb is excellent for classical music, pop, and more traditional genres. I would recommend using it on a whole mix to glue things together or on a buss to glue a group of tracks together or put them in the same space.

> TASK – Implement a hall reverb into a mix. Experiment with where you use it, i.e., try it on individual components, groups, and the whole mix. How does how you implement it alter how it affects your mix's clarity?

Day 5 – Different types of reverb: Chamber

Next on the list is chamber reverb. Historically, studios would build their own reverb chamber, which consisted of a speaker and a microphone placed inside a reflective room. The track would be sent out of the speaker, recaptured by the microphone, and routed back to the recording console. Given that the construction of every chamber is different, each has its own characteristics. For example, the Abbey Road Studios chamber is legendary and can be heard all over The Beatles' records.

Chamber reverbs are similar to halls in many ways in that they deliver a lush, ambient-soaked sound. But unlike hall reverbs, they offer an additional sense of clarity you don't get from the washed-out feeling of a hall reverb.

Chamber reverbs generally sound pretty neutral, so they work well on all sorts of sounds, especially vocals, strings, and acoustic guitar. They're perfect for a John Bonham-like drum sound too. They're especially great on small ensembles and classical music.

> TASK – Take the same mix you used for your hall reverb experiment. Swap the hall out for a chamber reverb. How do their characteristics compare? Does one work better than the other in some places?

Day 6 – Different types of reverb: Room

Rooms can be any size, so how large is the room in your room reverb? Think of it as a mid to large-sized room in a house rather than a concert hall or recording studio. They're a smaller acoustic space than hall or chamber reverbs and aim to reproduce most accurately the sort of ambience you're used to hearing in the real world.

The average room tends to have flat, parallel walls. These reflective surfaces tend to cause sonic anomalies like standing waves, flutter echoes, resonances, and ringing. Don't worry too much about what these things are. What's important to know is that you would expect a room reverb to colour your sound imperfectly, unlike hall reverbs. These imperfections are very desirable as they add character and interest. They also tend to be quite energetic and lively due to their smaller size. Of all the reverb types, they're the most rock 'n' roll-sounding. They're also the easiest to slot into a mix subtly as their ambience is what you're most used to hearing in daily life, so it draws the least amount of attention to itself.

Room reverbs are versatile, sounding great on pretty much any instrument or voice. Use them with restraint to add a natural feel to a mix whilst keeping an intimate feel that sounds closer than what you can achieve from a hall or chamber reverb.

> TASK – Same task as before, but insert your room reverb this time. This will feel quite different. You'll definitely prefer it in some instances, but not in all.

Day 7 – Different types of reverb: Plate

Plate reverbs don't imitate a real acoustic space like the types we've explored so far. Instead, they artificially create reverberation by using a magnetic driver to drive vibrations into a large metal sheet. A contact microphone collects the vibrations of the plate. The resultant reverb is rich, dense, warm, and luscious. In the '80s, digital recreations became a big thing, with many guitarists investing in racks for their rigs.

There are a few significant differences between plate reverbs and all the others we've looked at. First, they exist in a two-dimensional space rather than a three-dimensional one. This means that the echoes have the same density throughout the reverb tail, whereas in a three-dimensional space, the echo density increases as the reverb tails out. Secondly, sound moves much faster through metal than it does through air. These factors give plate reverbs their signature smooth reverb tails. Finally, higher frequencies expel their energy more quickly than lower ones. This means that higher frequencies will live at the front of reverb tails, with lower ones tailing out later as they take longer to build up. These factors mean that plate reverbs sound shiny!

The most common applications for plate reverbs are on vocals and snare drums, but you can make your track stand out from the crowd if used well.

> TASK – Use a plate reverb on your vocals and snare drum. Do you like the shiny quality it brings? Compare it with something larger, like a hall. Consider circumstances in which the different reverb types may be more/less suited.

Day 8 – Different types of reverb: Spring

Spring reverb was first found inside Hammond organs in the '60s. Given that it can fit inside an instrument, you can safely assume it's small. And it is! Have you ever kicked or slightly dropped a guitar amp and heard a rattling, springy sound? That's the spring reverb jangling around inside.

The mechanics are simple. The audio signal is sent to one end of a spring (or several springs). This creates waves that pass down the spring(s). At the other end, the motion is turned back into an electrical signal, which is combined with the dry sound. But when a wave reaches the end of the spring, not all of it is turned back into sound. Some of it bounces back down the spring. It's these reflections that create the characteristic spring reverb sound.

The sound is clean and is perfect for creating vintage guitar tones. It's often described as bouncy. They tend to be more low-end-focused as the lower frequencies have longer wavelengths and thus bounce back and forth more easily along the springs. Spring reverb should be considered less like an acoustic space and more as an effect. In this way, it can be implemented to complement acoustic verbs.

> TASK – Experiment with a spring reverb on some electric guitar channels. Can you recreate a vintage vibe? Does it work in combination with another reverb type, or is it better in isolation?

Day 9 – Different types of reverb: Convolution

Convolution reverb is an entirely different beast because it can sound like anything. Some of the explanations I've read online are perplexingly complicated. But they don't need to be. Convolution reverb is simply the capturing of the acoustic qualities of any space. Hence, it can sound like anywhere, or rather, anything. To capture a space's sonic characteristics, you need to create what's called an impulse response. This is typically done by playing a sweeping sine wave up the audible frequency spectrum. Microphones record this, along with the resulting reverb, which allows us to see how the resulting audio has been affected by the space's acoustic character. The original sine wave is then edited out of the recording, leaving just the resulting features of the space. These impulse responses are then loaded into your convolution reverb, allowing you to recreate any space so long as you can get your hands on an impulse response!

Generally, the outcome of convolution reverb is more of a textural and timbral one rather than an ambient one. For this reason, it can be great for adding life to bland or lifeless tracks.

Note how I said that they can sound like any*thing* rather than any*where*. You can capture the impulse response of any space. For example, I have IRs from the inside of some beautiful Gibson acoustic guitars. I use these frequently on my cheap acoustics at home to make them sound much nicer and more expensive! You can get IRs for some bizarre spaces: The inside of car boots, parking lots, bathrooms, etc.[3]

> TASK – Do a bit of digging for IRs online. Can you find any quirky (and preferably free!) IRs to load into your convolution reverb?

Day 10 – Different types of reverb: Gated

Gated reverb isn't a type of reverb in its own right. But I love it so much as an effect that I had to talk about it. It's the combination of a reverb that is then gated to cut off the reverb tail. It was used to excess in the '80s especially. Think Phil Collins in Genesis, and you'll know what I'm talking about.

There are a couple of things that make gated reverbs especially effective. First, they work particularly well with larger reverbs. I like using chambers, but experiment to find something you like. Secondly, they benefit from having a reasonably consistent input signal. This is because you're going to set a threshold so that when the signal dips below it, the reverb will be cut off or gated. If the signal coming into the gate is inconsistent, then the gated effect will also be inconsistent, which is probably not what you're after. Options to deal with this are to ensure that the level of the sound you're applying the gated verb to is even, either by compressing it or adjusting velocities if it's a drum part. Alternatively, you can compress the reverb itself before the gate so that the reverb's signal is more even before being gated.

Once your reverb gets to the gate, it's just a matter of adjusting the threshold, attack, and release as you would on a compressor. A medium attack will allow the transient of the original sound source through before the reverb blooms. But the critical feature is to have a quick release so that the reverb tail is cut off dramatically. That will give you the classic gated reverb effect.

> TASK – Set up a gated reverb in a mix. Use it on a snare drum. Can you recreate a vintage '80s feel?

Day 11 – Different types of reverb: Honourable mentions

All the main types of reverb have now been covered, but there are a few variations that some people would add to the list. I'll run through them briefly so you're aware of them.

First, you have cathedral or church reverb. This is a type of hall reverb, but it's longer and wetter. The characteristics are based less on the size of the venue and more on the highly reflective nature of a cathedral's construction and its square shape. This leads to a longer decay time, sometimes up to as much as ten seconds.

Secondly, you have ambience reverb. This is a kind of room reverb. It has an extremely short decay length of around half a second. It primarily consists of early reflections, which are true to the original sound. It adds almost no colour or additional character. For these reasons, it's excellent on busses for subtly gluing groups of instruments together, locking them into the same acoustic environment.

Shimmer reverb is next up, which is just a pitch-shifted reverb. It's typically put up one octave. This effect was heard a lot in '90s dance music. It's used a lot on synth pads, string pizzicato, etc. It brings a lot of shine and sparkle to your sound.

Reverse reverb is precisely what it sounds like. You bounce down your reverb tail, reverse it so that it increases in level rather than decreases, and then place it in front of the original sound source to act as a sort of riser. It was traditionally achieved by physically playing a tape backwards and then recording it. It's used a lot in horror, fantasy, and sci-fi, but you're probably more familiar with it at the drop of some of your favourite dance tracks.

Finally, you've got non-linear reverb. A normal reverb tail is linear (in actual fact, it's not; it's logarithmic, but let's keep things simple). Any reverb that is altered to decay differently is therefore non-linear. The science behind the sound is a bit involved, but they mostly all end up sounding much like a gated verb, so let's lump them together, shall we?

> TASK – Do any of these alternative reverb types appear in your DAW? If not, try to recreate them. Shimmer is my favourite, so give that a go.

Day 12 – Reverb parameters: Type, size, decay, and mix

Now that you're well informed on all the different types of reverb, you should be well placed to start selecting them in an informed manner. That's the first half of the battle won. The second half is to understand what all the controls do on a reverb plugin and what you need to tweak to dial in the perfect sound. Let's cover this now.

Lots of reverb plugins cover multiple different reverb types. You can now choose your reverb type in an informed way.

The size control is next. Don't let this worry you. When you choose the type of reverb you want, it may well alter its size relative to the type. You're unlikely to want a tiny hall or an enormous room, so these parameters are often linked. All you need to do is dial in exactly how large or small you want your space to be.

Decay follows this and will be stated in seconds or milliseconds. This is the length of time it takes for your reverb tail to disappear and is probably the most important control on the plugin. Too short, and it will sound unnatural; too long, and it will start getting in the way and cluttering up your mix. We'll talk about this more later, but as a rule, you usually want the decay to cooperate with the phrasing of the music. So, look for the reverb tail to have just disappeared in between musical phrases or hits. It's a bit like a compressor's release in this sense.

You're also likely to have a mix control, which may be labelled as dry and wet. If your reverb has been set up on an aux send, you'll want to turn the dry to 0% and the wet to 100%, so you only hear the reverb and not the original sound. If you're using the reverb as an insert on an individual channel, you'll want to play with this mix control to taste.

> TASK – Study your DAW's stock reverb plugins. Can you identify the type, size, decay, and mix on all of them?

Day 13 – Reverb parameters: Pre-delay, early reflections, and diffusion

Pre-delay is where it begins to get more involved and where most people get lost. Pre-delay is the length of time it takes for the signal to get to the boundary of the space and return to the listener. So, it's the time it takes for the reverb to begin. It gives you an indication of the size of the space, with longer pre-delays implying a larger area and vice versa. For pre-delay to sound natural, a length of 50 milliseconds or less is required. But longer times can still be used if, for example, you're trying to keep things forward in the mix. You'll often be looking for a pre-delay time that isn't so short that it muddies up and blurs the transients of the sound source but isn't so long that it is obvious. Lots of plugins will allow you to tempo-sync your pre-delay with your project tempo.

Early reflections are sometimes called initial reflections or pre-echoes. This control dictates the level of the first reflections you hear. This level is independent of the rest of the reverb tail, which is an important feature. Plate and spring reverbs won't have this option, as their reverb sounds are too small. Generally, the early reflections will be less than 100ms. Note that if this time is shortened to 40ms or less, your brain cannot discern it as its own sound but rather it will lump it in as part of the original sound. This is called the Haas effect.[4] The level of the early reflections will also indicate the size of the room, with louder ones suggesting a smaller space because the reflections have had less distance to travel. A common trick is to add the early reflections alone to the dry signal, which will enliven the sound in a pleasingly transparent fashion.

Finally, you have the diffusion control, which describes the scattering of sound. If you've ever seen what looks like a New York skyline scene stuck on someone's wall or ceiling, this is an acoustic diffuser. Its job is to scatter, or diffuse, the sound in every direction. This makes the frequency response more pleasant. The room's material and shape will affect the diffusion, with brick walls in an irregularly shaped room providing better diffusion than metal walls in a uniformly square room. Greater diffusion results in greater complexity in the reflection pattern, dramatically altering the reverb's overall sound.

Now you know what all the controls on your plugin do, you can begin to grab them with confidence, understanding what characteristic of the reverb you are affecting.

> TASK – Identify any remaining controls on your reverb plugins. Ensure you know what everything does. Reread any necessary information to ensure you fully understand the more complex parameters.

Day 14 – Eight steps to perfect reverb: Aux send or insert slot?

I'm now going to run through some crucial steps that will assist you in getting your reverb sound just right.

The first big argument people have is whether reverbs should be set up as inserts on individual channels or on their own aux sends.

Traditionally, reverbs would always be set up on aux channels. The obvious advantage here is that you can send multiple different sound sources through the same reverb in different quantities. This puts these sounds in the same acoustic space whilst simultaneously saving processing power. The difficulty comes when using multiple reverbs within the same project. If you have seven or eight different reverbs each set up on aux tracks but that are only fed by one channel, this wouldn't be a sensible use of your DAW's real estate. In these cases, placing the reverb on the channel itself as an insert makes much more sense.

A good example would be if you were only using a reverb on a kick drum, for example, for two bars of the song. You'd be much better off automating the reverb mix in this instance. Another example would be if you were using early reflections to enrich a mono signal.

In other words, there isn't a hard and fast rule. Use your brain. If a reverb is only for one sound, keep it on the individual channel. If it's affecting multiple sounds, use an aux send.

> TASK – Review a previous mix. Look at how you've set up your reverbs. Are you using aux sends, inserts, or a combination of the two? Are there more efficient ways to set them up?

Day 15 – Eight steps to perfect reverb: Selecting reverb type

Many people allow themselves to become entirely baffled by the process of selecting their reverb type. Usually, this is challenging because they don't understand the characteristics of the reverb types, so they don't know what they're listening for. We've addressed that problem earlier in the unit, which is half the battle here. To assist further, ask yourself these questions:

- What is the ideal space you want your sound in?
- What other qualities will add cohesion to your mix or sound?

The type of room you select should cooperate with the music you're making. A classical orchestra would benefit from a large hall, whilst a small room would be best in a heavy metal track. Notice how I've explicitly referred to room types here rather than reverbs as a whole. Reverbs are generally considered in two categories: Acoustic spaces, such as halls, chambers, and rooms, and mechanical effects, such as plate and spring. I find it helps to keep the two categories separate from each other. Use the acoustic spaces to give things a location and glue things together; use the mechanical reverbs as effects to add interest and character.

> TASK – Review some previous mixes. Can you find any instances where you've used an acoustic space where a mechanical effect may have been better suited? Or vice versa? Make the relevant adjustment.

Day 16 – Eight steps to perfect reverb: Set your size

Lots of people won't bother to alter the size of the reverb at all. They'll just pull up a preset, and away they go. But the size is key to ensuring that it beds into your mix cooperatively. The size should be relative to the mix density. Denser mixes will suffer significantly if larger-sized reverbs are employed, as they will inevitably cause masking and clutter. Meanwhile, sparse mixes may shine as the longer tails may help fill some of the gaps in the mix.

It's always worth remembering that a reverb that sounds spectacular in isolation may sound awful in the context of your mix. I've mentioned before the importance of making mix decisions in context and not in isolation, and this same principle matters with reverb.

The size will also affect the stereo image of the sound, with larger rooms providing a wider stereo image. In other words, the size allows you to scale from expansive reverbs down to intimate, narrow ones.

But the size needs to be considered in tandem with the decay time.

> TASK – Again, reviewing previous mixes, can you find instances where your reverb size is inappropriate for the song? Make the relevant adjustment.

Day 17 – Eight steps to perfect reverb: Set your decay

This is arguably the most crucial setting of your reverb because it denotes how long it takes for the reverb to die out. The most frequent mistake is to make it too long, which results in a muddy mix. When a reverb decay is too long, the reflections begin to smudge together, making it impossible to tell the difference between sounds. This is an obvious sign that your decay is too long.

As you would imagine, the decay time and room size are intrinsically linked in that you'd expect the decay time to increase as you increase the room size and vice versa. Most plugins will do this for you. However, you can force your reverb to oppose this traditional stance. It will sound weird, but some fabulously creative sounds are to be had by experimenting in this way.

There are a couple of common associations to be aware of: Longer reverbs provide a 'heavy' ambience while shorter ones are 'tight'. More extended reverbs also tend to be louder, contributing to the masking issues they come with.

Try to think of your reverb as an additional rhythmic element in your mix. If you want to avoid masking, then you should consider having it decay in between phrases. On percussive elements especially, you may wish the reverb on the snare to decay before the next kick. Or, on your vocal, you may want the verb to decay before the entrance of the following phrase. Your decay time will be affected by the track's tempo and the part's rhythm. There will never be one size that fits all here. So, yes, use presets as a starting point, but always dial them in to suit your track.

> TASK – Explore the relationship between size and decay. Can you create some unusual effects by forcing a large size with a short decay and vice versa?

Day 18 – Eight steps to perfect reverb: Set your pre-delay

Dialling in your pre-delay is a critical component in keeping your mix tidy and uncluttered. Pre-delay provides a little breathing space between your dry sound and your reverb. Very short pre-delay times give you next to no space between your dry and wet signal, which will indicate a small room, but this can also make your reverb more challenging to manage. Using a longer pre-delay time will give you a little breathing space between the dry and wet signal simulating a larger space, but importantly removing muddiness from your reverb.

Generally, you should look to dial in a pre-delay time that is long enough to provide clarity to the dry sound but not so long that it is noticeably audible. For this reason, although many reverb plugins will come with a tempo-sync option on the pre-delay, I recommend not using this feature. Instead, leave it un-synced and adjust the pre-delay length manually. This will allow you to pay close attention to the phrasing of the part you are adding the reverb to rather than assuming that a quantised value will do the job.

> TASK – Explore different pre-delay times. Can you find the sweet spot between separating the pre-delay from the transient without it feeling disconnected? Bear in mind this will change depending upon how transient-heavy the part is.

Day 19 – Eight steps to perfect reverb: Set your early reflections

Early reflections are often considered more as echoes or delays than reverb. This is a pretty helpful way to think of them. So, if you're looking to implement a gentle echoing or doubling effect without setting up a whole delay plugin, increase the level of your early reflections.

Proceed with caution here, though. If your early reflection time isn't set carefully, it may end up interfering with the rhythm and groove of your track. In this regard, lower early reflection levels will help with blending. But that's not to say you can't go with a prominent early reflection. Just be aware of the damage it may do if you're not careful with it.

The length of the pre-delay will make the early reflections clearer, so adjust these parameters in conjunction with each other to achieve what you desire. If you find that your reverb is becoming too dominant and difficult to manage, reducing or eliminating early reflections is an excellent place to start as they can cause timbral distortion and comb-filtering. Comb-filtering in this context occurs when two almost identical sounds play very closely, one being the dry sound and the other being the early reflection. When some parts of the sound are in phase and some are out of phase, parts of the sound will cancel each other, and other parts won't. This is comb-filtering.[5]

> TASK – Explore the relationship between pre-delay and early reflections. How do these parameters interact with each other?

Day 20 – Eight steps to perfect reverb: Set your diffusion level

Think of diffusion as texture. Increasing or decreasing your diffusion amount is like increasing or decreasing the complexity of the reverb's environment. Consider the increase in terms of people in a venue, objects in the room, and irregularities in the room shape, and you'll visualise it easily. The difference between an empty, cube-shaped venue and one full of people is night and day in terms of how the sound behaves in that space. More textured areas will provide warmer and smoother reverbs whilst less textured spaces will be flatter and more reflective, therefore providing clearer and brighter reverb.

Diffusion is also closely linked to decay time. If you like the length of your reverb but want to smooth it out, then increasing the diffusion amount is an excellent place to look. If you find that your reverb is making your sound too metallic or clangy, then diffusion is, again, an ideal place to look to curb this.

> TASK – Explore different diffusion settings to shape the tone of your reverb. Bear in mind that, with some plugins, diffusion will be linked to CPU usage. More diffusion will result in a greater CPU draw.

Day 21 – Eight steps to perfect reverb: Adjust the dry/wet/send amount

Finally, you'll want to adjust the amount of reverb present in your mix. How you do this will depend on how you've set the reverb up. If it's an insert on your channel, you'll blend the dry and wet amounts or use the mix knob, depending on how it's labelled. If you're set up on an aux send, you'll adjust the send amount, thereby adjusting how much dry signal is being sent to that aux channel. The reverb itself should be 100% wet.

The level of reverb you have in your mix will determine how your reverb interacts with the elements in your mix. How much reverb you need will depend entirely on your mix density and how prominent you want the reverb to be overall. With this in mind, dial in the reverb amount in the context of the mix rather than in solo.

When your tracks are mastered, reverb tails will often become more noticeable. Therefore, it's important to use reference tracks when adjusting your wet/dry balance to ensure you get a solid amount without overdoing it.

> TASK – Practise adjusting all reverb parameters together and setting your reverb levels in the context of the mix. Use reference tracks to inform your process.

Day 22 – Top tips: Less is more

Allow me to give you some of my top tips for getting your reverb just right in your mix.

Developing this issue of adjusting your wet/dry balance, the million-dollar question for many producers is, 'How much reverb is too much reverb?' I think there are two angles from which you can approach this question.

The first is from the point of view of reverb as an effect. If your reverb is there to serve a specific creative purpose in your track, then you'll want to be able to hear it in your mix. There's no point in doing something creative that sounds cool if it's then buried under a heap of other things. So, as a rule, for reverbs that are effects, ensure they can be heard.

The second angle is of reverbs that indicate the location of the performance. This is often where people get it wrong. Less experienced producers like the sound of their ambient reverbs so much that they overemphasise them in their mix. This has the effect of washing out the mix, muddying it up and losing the definition of crucial components. My advice for getting your ambient reverb level just right is this: Increase your send amount until it's just audible in the mix, and then back it off slightly. When you A/B your track with and without the reverb send, you will notice its absence, but it won't be so apparent that it's a distraction.

> TASK – Revisit some mixes. Review the level of your reverbs. Are your reverb effects audible? Are your ambience reverbs too loud? These are the common issues to address.

Day 23 – Top tips: Mono, stereo, or panned?

In the same way that you can have mono or stereo instruments, and you can pan these wherever you like, so too you can with your reverbs. There are some things to be aware of in each circumstance.

Mono reverbs are great for keeping your mix tidy as you can place them in a position in your stereo field. By moving the reverb slightly away from the dry source rather than having it right behind it, you will avoid masking early reflections. Whilst this can sound great as an effect, it won't sound especially realistic. The biggest problem with mono reverbs is that they don't resemble natural reverbs in any way, as natural reverbs arrive at your ears from all directions, i.e., in stereo. So, use a mono reverb if you don't desire the stereo width that a stereo reverb will provide.

Stereo reverbs will have different content on the left and right channels. Because of this, the stereo content of your reverb is likely to have significant phasing issues if you solo your reverb channel, meaning that the reverb in isolation does terrible things to your mono compatibility. This is another crucial reason to keep your ambient reverbs in check in your mix, as pushing them too much is likely to be detrimental to your mono compatibility.

You can pan mono reverbs and place them wherever you want in your stereo field. But you can also experiment with the panning of stereo reverbs. Before doing this, though, I would consider playing with the overall width of your reverb. By narrowing the stereo image of your reverb, you can create space at the extremities of your stereo field. This can be helpful for double-tracked electric guitars, for example. If you've got your rhythm guitar panned hard left and right in your mix with both going to a stereo reverb, then the reverb will sit right behind the guitars. If you narrow the reverb's width, you'll effectively bring the reverb closer to the centre and leave the guitars out wide, allowing them to maintain their punch and clarity.

> TASK – Explore mono and stereo reverbs, as well as panning positions. Review what each of them does to your mono compatibility.

Day 24 – Top tips: Pre- or post-fader?

If your reverb is placed as an insert it will be pre-fader. However, if you're using a reverb send, you have options. Namely, whether to use a pre- or post-fader send. Which one is right for you depends on your desired outcome.

By default, your send is likely to have been set up as post-fader. This means that whatever adjustments you make to your fader will be directly linked to how much signal is being sent to your aux channel. More often than not, this is precisely what you'll be after. You'll want the reverb level to be directly proportionate to the dry signal level coming from the channel. So, if you turn the channel up or down, the reverb level will go up or down with it. This doesn't mean the amount of send will physically increase or decrease. But with a post-fader send, the send amount comes after the fader, meaning any fader alteration will be reflected in the send amounts.

The opposite can be said of a pre-fader send. With a pre-fader send, the amount of signal being sent to the aux channel happens before the fader, which means that the level of reverb and the level of dry signal become independent of each other. In this way, you can choose to reduce the dry signal on the channel to very little or next to nothing, and the reverb level you've sent will remain the same, or vice versa. You can create some cool effects in this way, so it's worth experimenting.

> TASK – Experiment with pre- and post-fader sends. Ensure you understand how to change your sends to pre/post-fader, the difference between the two, and what flexibilities they give you.

Day 25 – Top tips: Treat reverb like an instrument

Reverb should be treated like any other instrument in your mix. This seems like an obvious statement to make, but it's so often overlooked. I would recommend that you be even more attentive to how your reverb interacts with your other mix elements above everything else. This is because it has the power to make or break your mix. Dialled in appropriately, it will enhance every aspect of your mix. Done poorly, it will leave you with a muddy, washy, undefined mess.

It's all well and good to say that you should pay it close attention, but what are you actually looking for? An obvious place to look is in the low end. By now, you're probably well used to high-passing things to keep the low end of your mix tidy. Your reverb is no different. High-passing the reverbs will keep the low frequencies clear and defined, keeping your track well grounded.

You may also look to low-pass. Reverbs can often have a lot of high-end content. This can become too much if not kept in check, making your mix bright and brittle-sounding. So low-passing or using a high shelf to cut upper frequencies is an excellent place to look.

You also want to ensure you're not overloading specific frequency points too much. For example, you may have a vocal with plenty of 3kHz in it, which you like. But by sending this to your reverb, you end up overdoing this frequency point. Consider notching some of this frequency out to keep that area in check.

I love to use a dynamic EQ side-chained to something else on the reverb send. Let's work with the previous example. Rather than just scooping 3kHz altogether on my reverb send, I can place a dynamic EQ after my reverb and set the sidechain input on that 3kHz notch to the vocal. In this way, when the lead vocal is present, that frequency will be notched out of the reverb but will bloom after the vocal exits so that it can still be heard in the reverb tail.

> TASK – Practise EQ'ing your reverbs. Low-pass and high-pass filters are good starting points. And then think about not overloading any particular frequency point.

Day 26 – Top tips: How many reverbs should you use?

This is the million-dollar question that so many ask. I like to think of it this way: How many locations do I want it to sound like these musicians are playing in? Typically, the answer is one! I don't want it to sound like I've got a drummer playing in a cathedral, a singer in a bathroom, and a guitarist in a cave. I want them to sound like they're playing together in the same space, even if they weren't when they were recording. So, my general advice is to stick to one reverb send for your acoustic reverbs (halls, chambers, rooms). The main reason for this is to avoid something called ambience collision. This occurs when you have different spaces colliding with each other that don't sonically make sense. What you usually want from your reverb is for it to be convincing, so sticking to one is a good way of achieving this.

I caveat this by saying that I don't think the same applies to reverbs used as effects such as plates and springs. To an extent, you can have as many of these as you like, although clearly, the more you use, the trickier it will become to manage how they all interact. In this instance, I refer you to yesterday's tip about treating all reverbs as another instrument in your mix. Just be sure to maintain awareness of how each reverb interacts with other mix elements, and you should be fine.

> TASK – Practise setting up a mix using just one ambient reverb. How does this limit/focus your workflow? Does it make your mix more cohesive?

Day 27 – The formula

Many producers look for a shortcut, a quick fix that will answer all of their reverb questions. The answer is 60,000 divided by your BPM. Simple! Or is it? This sum supposedly gives you the perfect decay time for your track. So, if you plum in a BPM of 120 you get 60,000 ÷ 120 = 500ms or 0.5s. But what does this 500ms value relate to? People often describe this as a quarter note (or crotchet) reverb length, which it is. But what this doesn't do is to give you any indication of room size or pre-delay value. Not so helpful then.

Let's try and make this work for us a little better. Let's begin with the same equation of 60,000 ÷ BPM. This quarter note reverb length is very short in reverb terms and would feel like a tight ambient space. Working with 120 as our BPM, this quarter note length is 500ms. To get a small room feel, it's commonly accepted that you should double this figure to 1000ms/1s, which will give you a half note (or minim) reverb. For a large room, you should multiply your initial 500ms by 4, giving you a 2000ms/2s length (a whole note (or semibreve) reverb); for a hall reverb, multiply your original by 8, giving you a 4000ms/4s length (two-bar reverb). So:

Tight ambient = 60k ÷ BPM = quarter note
Small room = (60k ÷ BPM) × 2 = half note
Large room = (60k ÷ BPM) × 4 = whole note
Hall = (60k ÷ BPM) × 8 = two bars

But then, we haven't considered pre-delay in this equation at all. You should be considering your reverb length as a total of the pre-delay and the decay time. To work this out, you can multiply the millisecond answer you got from the previous equation by 0.0156 to give you your pre-delay time. For example, with our small-room reverb of (60k ÷ 120) x2 = 1000ms, you can take the

$$60k \div BPM = L$$

$$L \times 0.0156 = P$$

$$L - P = D$$

Figure 5.3 Displays the equations required to work out pre-delay and decay time.

Source: Image by Sam George.

1000ms and multiply it by 0.0156 = 15.6ms pre-delay. If you then subtract this from the 1000ms, you get your decay length of 984.4ms. Let's try and tie all this together. In the following equation L = reverb length, D = decay, and P = pre-delay.

This is for a tight ambient reverb. Substitute the first equation for the small- or large-room or hall equations, respectively.

If you can be bothered to commit all of that to memory, then good on you! Most don't but instead choose to use their musical intuition to dial in the reverb settings with sensitivity rather than mathematical rigidity.

TASK – Use the equations to find appropriate pre-delay and decay times for different room sizes using different BPMs.

Day 28 – Unit summary

When I began writing this unit, I intended to combine it with delay so that all time-based effects were rolled into one. I soon found out that there's so much to consider with reverb that it warranted a whole unit of its own. That's the approach I promised to take with you: Thorough and considered.

As a reminder, here's what we've covered in this unit:

- What reverb is, why you need it, and what it does
- All the different types of reverb: Hall, chamber, room, plate, spring, convolution, gated, and some honourable mentions, too
- Type, size, decay, mix, pre-delay, early reflections, and diffusion
- Aux sends vs inserts
- Setting size, decay, pre-delay, early reflections, diffusion level, and wet/dry balance

- Setting the perfect amount of reverb
- Mono, stereo, and panned verbs
- Pre- and post-fader sends
- Treating reverb like an instrument
- How many reverbs to use in a mix
- The magic reverb formula

Well done for making it through all of that. As you've probably figured out by now, there are an awful lot of things to consider under every topic of music-production. This is why it takes so long to become an 'expert' (however you judge that) in this field. It's all well and good talking about these things in black and white in a book, but for them to make sense, you need to use them, explore them, and experiment with them in your projects. So, go and implement all of your newly acquired knowledge of reverb!

Checklist

- Do you fully understand what reverb is, what it does, and why you need it?
- Are you familiar with all the different reverb types?
- Are you confident in selecting an appropriate reverb type for your required application?
- Are you confident in adjusting all the reverb parameters, including pre-delay, early reflections, and diffusion?
- Do you understand the difference in setting up a reverb on an aux send vs as an insert?
- Do you understand the difference between pre- and post-fader sends?
- Do you understand the importance of pre-delay in maintaining clarity?
- Are you confident in setting appropriate reverb levels in your mix?
- Are you confident in EQ'ing reverbs?
- Have you explored mono and stereo reverbs?

You should move on to the following chapter only once you can answer yes to all these questions.

Further reading

1 Upton, E. (2013). *Why the Epidaurus theatre has such amazing acoustics.* [online] gizmodo.com. Available at https://gizmodo.com/why-the-epidaurus-theatre-has-such-amazing-acoustics-1484771888 [Accessed 9 Nov. 2022].
2 Jackets, T. (2021). *Sabine's formula and the birth of modern architectural acoustics.* [online] thermaxxjackets.com. Available at https://blog.thermaxxjackets.com/sabines-formula-the-birth-of-modern-architectural-acoustics [Accessed 9 Nov. 2022].
3 Virostek, P. (2019). *The quick and easy way to create impulse responses.* [online] creativefieldrecording.com. Available at www.creativefieldrecording.com/2014/03/19/the-quick-easy-way-to-create-impulse-responses/ [Accessed 9 Nov. 2022].

4 McAllister, M. (2021). *Haas effect: What is it and how it's used.* [online] producelikeapro.com. Available at https://producelikeapro.com/blog/haas-effect/ [Accessed 9 Nov. 2022].
5 Fuston, L. (2022). *Comb filtering: What is it and why does it matter?* [online] sweetwater.com. Available at www.sweetwater.com/insync/what-is-it-comb-filtering/#:~:text=Eliminate%20Comb%20Filtering-,What%20Is%20Comb%20Filtering%3F,%2C%20up%20to%2015ms%E2%80%9320ms [Accessed 9 Nov. 2022].

Unit 6

Delay and modulation effects

Day 1 – What is delay in music-production?

Delay is a potent tool in the world of music-production. It's used by musicians and producers the world over due to its versatility and coolness. But what exactly is it?

The common misconception is that delay and echo are the same. The two terms are often used interchangeably. And whilst there are strong similarities between the two, there are also some subtle differences that make them individual. Delay occurs when the sound source is captured, time-shifted (delayed) by a user-defined length of time, and then played back a user-defined number of times. Simply put, a sound is captured and stored and then played back when the producer says so, however many times the producer wants it. Echoes are also time-shifted repeats of the sound source, but they are designed to sound far more natural. So, you can expect an echo to have some filtering on the echoed signal and it will have a more natural tail. To be specific, echo is a subset of delay, but delay can be more than just echo. What more it can do is part of what we'll explore in this unit.

The technology that enabled delay to be created artificially originated in the '40s through tape loops. It was used primitively in the '50s but exploded in the '60s when artists like The Beatles made it a common feature of popular music. It further evolved through the '70s and '80s as technology developed.

As you'll find out throughout this unit, there are several different types of delay, and they all behave slightly differently, providing contrasting effects. Understanding the differences between these delay types and selecting and applying them creatively will separate you from other producers. It will provide your music with range and diversity, as well as a level of accuracy and precision that will make a massive overall difference to your productions.

Day 2 – Delay times

The human ear perceives different delay lengths in very different ways. We can generalise and say that short delays sound very different to long delays, with the primary difference between the two being whether or not the human ear can discern the original and the delayed signal as two distinct sounds. But the human ear is far more sensitive than that. We can divide delay times into five categories.

1. Very short delays of 0–20ms will produce a comb-filtering effect (assuming both signals are mixed in mono). You can refer to Unit 5.19 to remind yourself of what comb-filtering is. Delay times that are this short will alter the instrument's timbre, giving it a different character. However, if the dry and wet sounds are panned, this effect is known as the Haas effect.
2. 20–60ms delays are often considered to be doubling effects.
3. 60–100ms delays are perceived as two distinct sounds. This length of short delay is known as a slapback delay.
4. 100ms to quarter note delay is the most common, where echoes are distinguishable due to the length of time in between them.
5. Quarter note delays and above are sometimes known as Grand Canyon echoes.

There are different characteristics and delay types associated with each of these lengths, which I'll explore with you in this unit.

> TASK – As an initial exploration into delay, use your DAW's stock digital delay and dial in each of these five delay times. Try some blind tests by randomly dialling in a delay length and seeing if you can identify which category it falls into.

Day 3 – Primary delay types: Tape

There are three main types of delay as well as a load of subtypes. Each has its own set of characteristics and comes with its own pros and cons. The differences can be heard in the initial echo and the way the delay trails off.

First on the list is tape delay. It's logical to start here as tape delay is the earliest type of delay there is. The tape records the signal and then passes it through a playback head a little time afterwards. So, you hear the initial sound in real time and then the delay when it hits the playback head.

This system fell out of fashion due to the bulky and delicate nature of the tape machine. They're also expensive to purchase and maintain, so they are impractical for most musicians. Nevertheless, the sound of tape delay was made infamous by the likes of Brian May and Jimmy Page in the '70s, who used the EP-3 Echoplex. This sound is so desirable that you can now purchase plugin emulations of this specific tape-delay hardware.

Tape delay can generally be described as lo-fi but warm, with feedback that gets more and more distorted as it trails off. Tape isn't a precise medium, so you can expect to encounter pitch fluctuation and wobble. If you're after a vintage, gritty delay, then tape delay is your answer.

> TASK – If your DAW has a built-in tape-delay plugin, explore it. In particular, look out for controls such as wow and flutter. These are specific characteristics you get from tape that will contribute enormously to a vintage sound.

Day 4 – Primary delay types: Analogue

Analogue delay was invented to recreate some of the tape-delay sound without the hindering cost. They were designed to fit into guitarists' racks or pedals. They were meant to be portable and sturdy, allowing guitarists to get their sound on the road relatively inexpensively.

These pedals use a 'bucket brigade' circuit, which is a series of capacitors. Each capacitor takes a sample of the sound before passing the remaining signal onto the next, just like a real-life bucket brigade does with a bucket of water.

The resultant delay is similar to tape delay, but it is not the same. The most obvious difference is that bucket brigade analogue delay is darker than tape delay because it needs to be low-pass-filtered to retain sound quality. The heavily filtered sound is helpful but certainly isn't clean-sounding.

Bucket brigade circuits took a giant leap forward in the effects world as they could create much shorter delays than tape delays could.

> TASK – You're unlikely to have anything labelled 'bucket brigade' delay in your DAW. In truth, you may not even have anything labelled as analogue delay. Instead, you may need to watch some YouTube videos or purchase something third-party.

Day 5 – Primary delay types: Digital

Digital delay is the last primary delay type. Digital delays started as rack units in the '70s and '80s and are the cleanest of the delay types. Creating a delay of a sampled sound in the digital domain is pretty straightforward. The signal is only really affected by your sample rate and your analogue-to-digital conversion. Therefore, you can expect digital delays to sound clean, particularly in comparison to tape and analogue delays.

These days, though, loads of digital-delay plugins aim to recreate tape and bucket brigade delays. This makes them a sort of hybrid delay, bridging the gap between the analogue and digital worlds.

One of the main advantages of digital delay is the ability to tweak multiple characteristic features. You can often adjust how coloured or transparent and long or short the delay is. You may even get MIDI control. Some older digital-delay models don't have the best A/D and D/A converters, but modern ones generally offer 24-bit resolution. This is just something else to bear in mind.

> TASK – You certainly will have digital-delay options with your DAW. Explore the settings you have available. Do you have feedback? Cross-feed? Filters?

Day 6 – Delay subtypes: Slapback echo

Within these three primary delay types, there is a whole raft of subtypes that are more commonly referred to. Let's explore them now, one at a time.

First up is slapback echo. Slapback echo, or delay, is a single, short repeat of a sound. It first became well known as an effect on vocals and guitars in the '50s.

As a mixing device, it's excellent for drawing attention to aspects of a mix as the short delay time adds additional excitement and energy. It won't clutter your mix too much as the single delay doesn't have a long decay tail.

Typically, you'll find slapback in songs with faster tempos because of their energetic nature. You'll usually want to use a tape or analogue delay when producing slapback echo to get that vintage vibe associated with this effect.

If you're looking for some settings to help you dial in this sound quickly, here you go: Set your delay time somewhere between 75–200ms, make sure your feedback is off, put your mix somewhere above 50% and keep the modulation off, or just add a slither of pitch modulation subtly.

> TASK – Create a slapback echo. Try with both a tape-delay plugin and a digital delay. Which one's characteristics do you prefer?

Day 7 – Delay subtypes: Doubling echo

A doubling echo is meant to replicate the sound of a double-tracked recording. In this sense, what you're not trying to achieve here is a perceivable echo. What you're looking for is a doubled or thickened sound. You'll want to use a very short delay time, probably somewhere between 30–50ms.

It's important to point out here that, whilst this effect is supposed to do the job of a double-tracked part, it doesn't sound the same. It sounds noticeably different, which is evident when you consider that it is simply replicating the same audio signal rather than producing another performance with a human variation. With that being said, you can improve the legitimacy of this effect by dialling in some light modulation. Modulation will be covered thoroughly later in this unit.

Settings you'll be looking to dial in for doubling echoes are: A delay time of 30–50ms, mix at around 50%, and light modulation. You'll want some feedback here, the length of which will vary depending upon the input source, but lean towards a shorter amount.

> TASK – As before, experiment with creating doubling echoes using different delay plugins. Do you find you prefer the sound of one over another? What is it that you like?

Day 8 – Delay subtypes: Looping

As its name suggests, looping delay is a looper set up using some sort of delay. Originally, tape delays would be used to make this looped sound. You would capture a whole phrase and repeat it around and around in a tape loop, allowing you to layer additional sounds or harmonies on top.

This principle was developed in the '80s with digital-delay units, eventually resulting in the invention of the dedicated looper, which is used so prevalently today by artists such as Ed Sheeran. A looping delay is effectively a looper minus the dedicated record and playback controls. This means that the delay parameters, such as delay time and feedback control, will dictate aspects of your performance.

Looping delays are widespread in synthesis, especially in modular synths. They allow performers to layer sound upon sound, making rich and complex performances relatively easy.

Some guideline settings here are: Delay time between 500–4,000ms with a reasonably high feedback amount, mix around 50% with no modulation.

> TASK – Experiment with looping delays. They're great for creating unique textures that have a feeling of evolving over time.

Day 9 – Delay subtypes: Multitap

Multitap delay is a combination of two or more delays that you can arrange in many ways. Due to their complexity, they're almost always digital.

There are a vast range of ways in which the different taps can interact with each other. Consider a simple example with just two taps. The first tap is fed into the second, but only the second has feedback on it. Or perhaps rather than the taps being in series, they are in parallel, with the two different taps being combined on output. The more different taps you add into the mix, the more complex the outcomes you can create.

Multitap delay is commonly used to create reverb-esque effects. If you play around with the modulation and filtering of each tap, you can begin to simulate different types of diffusion. To make anything vaguely realistic-sounding, you'll need a lot of taps, but I would suggest that if you want it to sound like reverb, use reverb! If you want to create a cool reverb-like effect, however, experiment with your multitap delay.

> TASK – Often, with multitap delays, you can adjust multiple parameters per tap. In Logic's Delay Designer, you can adjust each tap's cutoff frequency, resonance, transposition, panning, and level. See what you can achieve in your DAW.

Day 10 – Delay subtypes: Ping-pong

Ping-pong delay is great fun, as it sounds as if the sound bounces from left to right and back again. There are a few different ways in which ping-pong delay can operate. Commonly, it will be fed solely into one side. This signal will then be fed back into the opposite side. The process will repeat to create the bouncing effect. Generally, each delay will decay more and more, so the sound gradually dies out.

You may also come across options to feed the signal into both sides simultaneously. But with the resultant feedback still alternating between left and right, you'll still get the ping-pong sound. You won't generally come across a designated ping-pong delay unit or plugin, though. It's usually a feature within a unit.

Ping-pong delay is excellent and supercool, but use it sparingly. It can quickly become tiresome. It works most effectively on simple, short sounds. Therefore, I recommend using it to accent specific moments or phrases. Don't stick it on a whole vocal line, but rather on one word you want to add weight to at the end of a line.

It's also used a lot in reggae and dub music, most notably on drums. Try it on your hi-hats to add additional interest to your groove and make your top end wider.

> TASK – Add ping-pong delay sparingly to a track. Try it in a few specific locations. For example, place it on individual words at the end of phrases, on a particular percussion hit, or on something else that requires emphasis.

Day 11 – Delay subtypes: Dub

Dub delay was explicitly invented to serve dub music. It's a reasonably complicated effect, usually containing multiple effects chained together, but there are common elements that will lead you towards that signature sound.

The roots of dub are in Jamaica in the '60s and '70s. Most dub artists were working with old equipment, basically referring to tape delays. The classic effect for dub and reggae is the Roland Space Echo, but most tape delays will do the job just fine.

What you're looking for is for the delay to become more and more distorted as the feedback progresses. Commonly these days, you'll find that the feedback is filtered gradually to make things sound as if they're falling away.

Dub delay sounds musical. You can get stuck in with the delay time and feedback amount without doing anything too damaging to your mix. You can also use pitch modulation to create some cool effects, but the magic happens when you modulate the delay time to make the classic tape-flutter sound.

As a quick aside, with all these delay types, you can use the previous calculations I outlined concerning reverb in Unit 5.27 to tempo-sync your delays too, if you so desire.

> TASK – There are some great emulations of the Roland Space Echo out there. Universal Audio, IK Multimedia, and Overloud all do one. If dub is within your wheelhouse, I recommend picking one up. If not, watch a few demonstrations on YouTube.

Day 12 – A reminder about phase

That's all the major players in the delay world covered, but other well-known effects are delay effects, too – namely chorus, flanger, and phaser. Before explaining how these work let's remind ourselves about phase and how it works as it's the crucial component in each of these effects.

As you know, audio is made up of sound waves. These waves are the result of vibrating air molecules. The vibration of air molecules is gathered by our eardrums, which is then interpreted by our brains as sound.

When two different sound waves interact, the resultant outcome depends on how those waves overlap. When crests of sound overlap, they combine to make a louder sound. When a crest and a trough overlap, they cancel each other out. These actions are known as constructive and destructive interference.

When two identical soundwaves line up perfectly, they are known to be 'in phase' with each other. If 100% destructive interference occurs, though, with crests of one and troughs of the other aligning perfectly, they are '180 degrees out of phase'.

You'll almost certainly not come across true destructive interference in your day-to-day life. So many external factors affect how sound moves and travels around you that the chances of encountering sound that is 180 degrees out of phase are very slim. However, in music-production, you can create exact replicas of sounds with the click of a mouse. This empowers you to experiment with the phase of sounds – a technique employed in the subsequent three effects we'll explore.

> TASK – Manually experiment with the phase of a sound. Duplicate an audio region. Flip its polarity and adjust its position left and right slightly to create some interesting sounds.

Day 13 – Modulated delay: Chorus

When your audio enters a chorus plugin, a copy, or multiple copies, of your audio will be created. This signal or signals are identical to the original. They are often called voices within a chorus plugin.

However, your new voices will be delayed very slightly. This will create a difference in phase between your original sound and your new voice(s).

Your delay time will then be modulated with an LFO (low-frequency oscillator). We'll talk more about LFOs in Unit 9, but for now, just think of them as something that will modulate your sound continuously. Modulating the delay time this way will gradually alter the wavelength of the resultant sound. This will result in varying the voice's pitch. You can then blend this sound back in with the original.

Chorus, therefore, affects the pitch and timing of your sound source. This effect simulates multiple singers or instruments who would never perform perfectly with each other but would be reasonably close.

Because your pitch is modulated in the additional voice(s) rather than the original sound, it's rarely at the same frequency as the original. This means that constructive and destructive interference is minimal.

You can use chorus to wash out sounds, making them feel more ambient. You can push the effect if you want it to sound prominent, but the result will lose some presence in your mix. For this reason, it can be great on textural layers and things that don't need to be at the front of your mix.

> TASK – Explore chorus within your DAW. The critical parameters are rate, intensity/depth, and the number of voices.

Day 14 – Modulated delay: Flanger

Flanger works very similarly to chorus. A copy of the original sound is created and delayed. However, the delay time will generally be much shorter than with chorus. An LFO is then applied to the duplicate signal, again modulating the delay time. This is then blended back in with the original. Because the two signals are identical and very closely aligned due to the short delay time, some resonances will be created at specific frequencies. But more notably, comb-filtering will occur.

This filtering occurs in both chorus and flanger, but the affected frequencies will be much higher with flanger due to the shorter delay time and, conversely, much lower in chorus.

As the LFO in the flanger alters the delay time, the filtering points will move. This is what creates the distinctive resonance sweep.

However, flangers will often feedback the output to the input. This will magnify the effect of the notches and resonances, contributing to the harsh, metallic timbre associated so typically with flanger.

The overall sound and impact of flanger is great but can quickly become tiring to your listener. Therefore, it's great to use during transitions as short-term ear candy rather than throughout a whole section of a song.

> TASK – Whereas chorus can be used subtly throughout a mix, flanger tends to be used more sparingly. Revisit a mix and identify one or two places where flanger can be implemented to create interest. In addition to the rate and intensity dials, the feedback option here is critical.

Day 15 – Modulated delay: Phaser

Phaser, unsurprisingly, works in a very similar way to both chorus and flanger, but with one key difference. Like its predecessors, the original signal is duplicated, but instead of delaying the copy, it is passed through an 'all-pass filter'. This sort of filter doesn't affect the audible frequency content as you'd expect of filters on an EQ. What it does is introduce a phase shift around a specific frequency point.

Placing another all-pass filter immediately after the first one will create one single notch such as you'd see on a comb filter. By chaining multiple all-pass filters one after the other, phasers generate a series of non-harmonically related notch filters.

Then an LFO is used to modulate the notch filters, creating a similar motion to that of a flanger. Because the notch-filter points are inharmonic, the effect is gentler than a flanger but more potent than a chorus.

In some phasers, you can choose how many notches are created. Sometimes these will be labelled as poles. You may also come across a feedback option, giving it similar capabilities to a flanger in this regard.

As you can now see, these three effects are all similar in that they contain phase-shifting in one way or another. Chorus combines it with pitch, flanger employs it to create harmonic comb filtering, and phaser uses all-pass filters to phase shift without any delay.

> TASK – A/B an audio sample with all three effects. Try to configure the main parameters (rate and intensity) equally so you can compare their sounds as fairly as possible. Do you have a favourite?

Day 16 – Top tips: Have an intention

Now that you're aware of all the different types of delay at your disposal, let me give you some top tips for getting the most out of them. As I go through these tips, you'll probably realise there are as many similarities in the approach to delay as there are to many other aspects of your mix.

First on the list is to have an intention. Put simply, think about what you're trying to achieve or how you want it to sound. Are you trying to nail the sound of a specific genre? Do you want to position the delay somewhere different to the source sound in your stereo field? Are you trying to push the sound back in the mix? By asking yourself the obvious questions about what you're aiming for, deciding how to get there will become much more manageable.

You can affect so many aspects of a sound with delay. You may focus on changing its front-to-back position or its left-to-right. You may use your delay to provide weight to something or to create ambience and depth. A slapback delay will be indicative of specific genres. How you alter the timing of your delay may reinforce or change the overall groove of your track.

When confronted with all these questions, the good news is that most delay plugins will come armed with a range of presets. These are often an excellent place to start. They will most likely be named with valuable things that indicate if it's most appropriate for a specific instrument, genre, or particular effect. As delay plugins will often be tempo-synced to your project, you can usually get away with calling up a suitable preset, blending the wet/dry and leaving it. But as you grow in confidence, you will undoubtedly want to delve deeper into the finer details so as to enhance your sound and sit things perfectly within your mix.

> TASK – Spend some time browsing the presets you have within your delay plugins. How are they presented? Are there any indications as to genre, tempo, rate, or complexity?

Day 17 – Top tips: Be selective

It's common when you're new to the world of music-production to fall into the trap of overdoing the amount of reverb you put on your mix. Equally common is to over-exaggerate your delay amount. Often, you'll come across both effects being overused simultaneously. Training yourself to understand how to dial in a sensitive amount of each of these effects without washing out your mixes is a skill that must be developed over a period, much like learning your frequency characteristics or learning to hear compression.

The biggest tip I can give you here is that it's OK to have tracks that are 100% dry. Don't feel that you must put delay, or indeed reverb, on every track in your session just because you have the aux send set up and ready to go. When considering whether something would benefit from some delay or not, you should always try to refer back to your intention: Are you aiming to simulate ambience, adding a sense of direction, evoking specific stylistic characteristics, or adding emphasis to something?

By being selective with what, where, and how much delay you put on things, you will directly benefit the overall clarity of your mix and the control you have over it. More often than not, this is precisely what you'll be wanting: Control and clarity.

> TASK – Create a delay 'sense check' list and have it somewhere visible. On it, ask if you're looking to add ambience, add a sense of direction, evoke a specific style, or add emphasis. If you don't want any of those things, you don't need delay.

Day 18 – Top tips: Work out your timing

Part of making your delay bed well into your mix is about working out the optimum timing to provide the impact you're after. Think of it like a well-timed joke, where the pause before the punchline is skilfully crafted for maximum impact. Any delay is, by its nature, rhythmic. It will therefore be influential in some way. How quickly or slowly the repeat of the initial sound is heard will directly influence how it is perceived.

I previously mentioned how you could use the same equations we used earlier in Unit 5.27 for reverb to work out delay timings. In this way, for example, you could set up a stereo delay on your drums timed perfectly to a quarter or half note and blend subtly underneath the dry signal.

But what if you didn't want something to lock perfectly into your track's tempo? What if you wanted something that was purposely 'off the grid' (not locked into your track's BPM) to stand out, be disorientating or add more complex interest? In this case, try looking towards a time that is a millisecond value of a prime number such as 23, 29, or 31. The benefit of using delay lengths that are prime numbers is quite scientific. Essentially it boils down to the fact that numbers divisible by two can generate a sense of tone, whereas prime numbers sound more neutral. For your reference, I've listed all the prime numbers up to 1,009 in Figure 6.1.

But proceed with caution here. If you're deliberately implementing a delay time that is not tempo-synced, always ensure that you check its impact by ear. Sometimes, regardless of the maths you've used to work something out, it will just sound wrong. Your ears won't lie to you. If you don't like how it sounds, tweak it.

> TASK – Explore some non-tempo-synced delay times utilising prime numbers. How do they differ in feel from delays that are locked to the grid?

**3 5 7 11 13 17 19 23 29 31 37 41 43 47
53 59 61 67 71 73 79 83 89 97 101 103
107 109 113 127 131 137 139 149 151 157
163 167 173 179 181 191 193 197 199 211
223 227 229 233 239 241 251 257
263 269 271 277 281 283 293 307 311
313 317 331 337 347 349 353 359 367
373 379 383 389 397 401 409 419 421
431 433 439 443 449 457 461 463 467
479 487 491 499 503 509 521 523 541
547 557 563 569 571 577 587 593 599
601 607 613 617 619 631 641 643 647
653 659 661 673 677 683 691 701 709
719 727 733 739 743 751 757 761 769
773 787 797 809 811 821 823 827
829 839 853 857 859 863 877 881
883 887 907 911 919 929 937 941 947
953 967 971 977 983 991 997 1,009**

Figure 6.1 A list of all the prime numbers between 0–1,009.
Source: Image by Sam George.

Day 19 – Top tips: Leakage

Imagine this scenario: You've recorded a band live rather than multitracking them. Perhaps it's a live studio recording or a recording of a gig. You come to mix the vocal and decide that you want to put some delay on it. But once you've implemented the delay you like, you hear that it is doing weird things to your mix. Upon closer inspection, you notice that your vocal mike has also captured some natural ambience from the recording environment, including some drum bleed. So not only are you adding delay to your vocal, but you're delaying whatever drums and other things have bled down this mike too. You must proceed very cautiously in this context to ensure you don't introduce any weird timing issues. You can quickly end up muddying your stereo image or negatively impacting your mix in other ways that will detract from the song.

Two great tricks can be used to assist with this problem. Option one is to filter the sound going into and/or coming out of the delay so that only the frequencies you want to be delayed or the parts of the delay you want to hear are audible.

The second option is to sidechain-compress the delay, with the input of the sidechain set to whatever the offending sound is. Using our previous example, if the snare drum was the offending sound that had bled down your vocal channel, you could set the sidechain input on your compressor to the snare channel to duck the delay each time the snare hits. Just make sure that you place the compressor after the delay in the signal chain so that the ducking occurs post-delay.

> TASK – If you've got any mixes of live recordings, check them to see if you've inadvertently added disorientating delays. Then, try the sidechain technique outlined previously. Don't fret if you can't get it working immediately. It's not easy!

Day 20 – Top tips: Processing delays

I've said it before when talking about reverb, and I'll repeat it now concerning delay: Treat it like another instrument in your mix. In the digital age of preset surfing, it is too easy to flick through until you find something that works well enough for you and move on without giving it a second thought. You may not have the confidence to dive into the delay plugin you're using to adjust the sound to your taste. But don't forget that you can always process the delay after the fact, with external EQ, further modulation, or additional dynamic processing.

In the previous tip, I talked about controlling the dynamics of your delay. EQ'ing your delay will help bed it into your mix, ensuring it contributes positively rather than emphasising muddy frequencies. Using a modulation effect such as chorus on your delay can help to get it out of the way of the initial source if you find they're mushing together too much.

I firmly believe you should get to know any plugin that you're using intimately. Only by knowing precisely what every control does will you get the most out of it. On a delay plugin, after the obvious delay time and wet/dry controls, two things are always worth identifying: The feedback (for how long the delay lasts) and the filters (usually high- and low-pass).

> TASK – If you still have controls on a delay plugin you're unfamiliar with, spend time getting to know them. You should know what everything does. Then, explore processing delay sends. You can use EQ, compression, or even other modulation or time-based effects such as chorus and reverb.

Day 21 – Top tips: Automation

Automation is a massive topic that I will be covering in detail in the next chapter of this course. But I felt it necessary to touch upon it briefly here in the context of delay.

There are so many fun, cool, and exciting ways that you can automate delay that many producers simply don't consider. Let me throw a few of them at you now. Try automating the rate so that it speeds up as it fades out. Or try the same thing but automating the feedback amount to increase as it fades out to elongate the decay tail. Automate the overall delay level to sculpt precisely how the delay fades out. If your delay plugin has a freeze control, then automating this is excellent for adding tension to build-ups. What about automating the filters so that you gradually low-pass the delay on a long decay tail rather than fading the overall level?

You can automate delays to create unique sounds in so many creative ways. Your only limitation here will be how well you know your plugins. I refer to yesterday's point about knowing your plugins intimately.

> TASK – I've just given you a bunch of exciting ways to automate your delay. I know we haven't covered automation yet but give these a go now. They're great fun!

Day 22 – Top tips: Create manual delays

Sometimes, something that sounds like a delay to the untrained ear may turn out not to be. Often, when you hear just a single repeat of a sound, it is created, not by dialling in a plugin to make the effect, but by manually copying the content you want to delay onto a new track and then adjusting it there. If you're familiar with vocal throws, you'll have an idea of what this sounds like.

There are a few benefits to creating manual delays rather than automating delay plugins. First, it can often save on your CPU consumption. The fewer things you're automating, the fewer your computer must process in real time, so the smoother your session is likely to run. Secondly, by having just the phrase or section of the sound you want to hear delayed copied onto a new track, you don't risk any additional sound creeping into your delay. This can happen either by not automating in the precise position required to feed the exact section in and out of your delay plugin or by latency introduced by a heavy session causing your automation to lag a little. Thirdly, by having the delayed content on its own track, you can EQ, pan, and render that content to your exact specification without worrying about what other content you may be affecting.

Whilst it may seem like the longer route to the same result, in practice, it is a time and CPU saver and can prevent hours of head-scratching when you can't work out what it was you automated that is now causing you additional unforeseen issues.

> TASK – Explore creating delays manually rather than using a plugin. Vocal throws are a great entry point into this. You'll find lots of examples on YouTube.

Day 23 – Top tips: Flamming delays

It is likely that you will be working with a mono signal at some point on your music-production journey and will want to make it sound wider. You may think using your DAW's stereo delay plugin will do the trick for you. Put it on, dial it in, and it will instantly sound full and wide. That is until you sum your mix to mono. I talked a lot earlier about mono compatibility, so you know its importance already. You may well hear some unnatural slapping like a bad flutter echo or even comb-filtering in this instance.

Don't fret though. The 'stereo delay on a mono source' trick is still a handy device so long as you know the following: All you need to do is manipulate one of the sides a bit. If you're working with a melodic part, a pitch-shifter that modulates just a little bit on one of the sides will be enough to prevent this nasty mono flamming. If you're working on a percussive mono element, then you can use a multiband transient shaper. By gently adjusting your different bands' sustain and attack, you will again avoid mono flamming. Just remember to use the transient shaper on your delay, not your original sound.

> TASK – Practise implementing stereo delays on mono sounds to create width. Use the techniques outlined previously to avoid flamming when summed to mono.

Day 24 – Top tips: Panning delays

There are a few different scenarios to cover when considering panning delays. Let's look at panning mono signals first.

Consider a lead vocal that is down the centre of your mix. To avoid any delay you place on your lead vocal interfering with the lead vocal itself, you will most likely want to pan it. But if you employ a mono delay on your lead vocal and pan it left or right, you will be pulling the lead vocal away from the centre of your mix. So instead, you will probably use a stereo delay to generate a delay on both sides of the stereo image and therefore keep your vocal nicely focused down the centre of your mix. If your delay plugin can designate the position of the delay, you can do so within the plugin. If not, you can use a direction mixer to narrow the stereo delays as much or as little as you like.

Now consider a backing vocal. This vocal is already panned to one side of your mix so as not to interfere with your lead vocal. You don't want to put a stereo delay on it as it doesn't need to sound wide like the lead vocal. Instead, you may use a mono delay. In this case, it is common practice to pan the mono delay opposite the source signal in the stereo field. This will provide an additional sense of width without detracting from anything else and will keep your stereo image nicely balanced.

What about if your sound source is a stereo signal to begin with? In this case, you should consider precisely what it is you want to achieve with your delay. The sound is already present on your mix's left and right sides, so a stereo delay is probably less effective. On the other hand, maybe a mono slap-back delay is what you're after, or perhaps a ping-pong delay that will bounce from side to side. Just remember to have an intention, and you'll be fine.

> TASK – Experiment with these techniques. Try stereo delays on mono sources, mono delays on mono sources that are already panned off centre, and other delay effects on stereo sources.

Day 25 – Top tips: Tempo-syncing

Your natural inclination when selecting your delay times is likely to be to tempo-sync them to your project. Especially if you are working within the electronic music world, this is likely to be your default setting. Delay is the most audibly distinct time-based effect. Any offbeat echo – in this context, I refer to offbeat as meaning not on the grid, rather than in the traditional sense of falling directly between the strong beats – will clash with the song's timing and cause rhythmic confusion. Any delay time that is longer than a quarter note is particularly susceptible to this. But that doesn't mean to say that your delays should always be tempo-synced.

Very short delay times, those that are 100ms or less, will probably not be perceived rhythmically within your track. So, their delay time will become more about feel than specific millisecond value.

In other contexts, a tempo-synced delay may not be audible due to the nature of the part it is applied to. For example, if you place your delay on a hi-hat playing an eighth note pattern, some echoes will be masked by the original source. You could try a triplet-value delay to work polyrhythmically with the original pattern, but you will still have overlap on your strong beats. An advantage of a non-tempo-synced delay in this context is that they can draw attention to a part precisely because they work against the natural rhythm of the song. I would use this sparingly to create tension by implementing these offbeat echoes that you can later resolve.

> TASK – Practise creating offbeat delays that establish rhythmic interest. Can you bed these into your tracks so that they enhance rather than detract from the overall groove?

Day 26 – Top tips: Making sounds bigger/wider

Making your track sound big and wide is often the aim of many producers, regardless of the genre they're working in. To achieve this, a single echo with a short delay time is the solution. You just want to ensure that it's not so short that it creates comb-filtering, nor so long that it is perceivable as two distinct sounds. The advantage of doing this is that you can stretch a monophonic signal wider across the stereo field to create a bigger image. Delaying this way can often lose a bit of focus, but this doesn't always matter.

The trick with making anything sound big or wide is that it will only sound big and wide if it has something to contrast with. For example, the tallest building in the world only looks so against the backdrop of all the other buildings around it that are less tall. So, if everything in your mix is big and wide all the time, its impact will be significantly lessened as it will have no context within which to sound so. However, if the big and wide sound comes after a smaller and narrower passage, you will maximise its potential impact.

In general, producing interesting music is all about creating contrast. If you can engrain this ethos into your workflow, you will find yourself producing creative mixes time and time again.

> TASK – Review the delay in a mix. Have you employed delays that run throughout the whole track? Could you automate these on/off in specific sections to provide width in certain places?

Day 27 – Top tips: Enhance important things

The principle of creating contrast can be developed to help you make critical features of a mix stand out. If you place a noticeable delay on your lead vocal throughout your whole chorus, yes, you will add weight to the vocal throughout that section. But every word will have equal importance placed upon it. Lyrically, this is unlikely to be the case. Not every word, or even every syllable, carries equal weight in their lyrical or melodic value.

Instead, I encourage you carefully to select the notes or syllables that you want to add value to and enhance these with a prominent delay. It may be the word at the end of a phrase where it has time to breathe in the mix. Or it could be just the hook that you will want to focus on. Maybe there are two or three significant melodic moments that you want to enhance. Whatever you add value to, by being selective rather than painting whole passages with the same brush, you will undoubtedly create more exciting and successful mixes.

You can apply this same selective process to any aspect of your mix. It could be a singular snare hit, a bend in the guitar solo you want to bring out, a bass slide or swoop, a horn stab, or a keys glissando. The applications are almost limitless.

> TASK – Can you find the most important parts of your song and add value by emphasising them with a delay? This style of delay can be more prominent than a generic delay that is used more as a textural device. Delay for emphasis should be heard clearly.

Day 28 – Unit summary

So, there we go! That's your time-based effects covered. And you now really are well on the way to beautiful mixes. To recap, in this unit, we've covered:

- What delay is, and different delay times
- Tape, analogue, and digital delay
- Slapback echo
- Doubling echo
- Looping delay
- Multitap delay
- Ping-pong delay
- Dub delay
- Modulated delay effects: Chorus, flanger, and phaser
- Having intention and working out your timing
- Leakage and processing delays
- Delay automation and creating manual delays
- Avoiding flamming delays
- Panning delays
- Tempo-syncing
- Making sounds bigger and wider
- Enhancing the important things

There has been a lot of information in this unit. The best way to become familiar with all these delay types and techniques is to go and experiment with them one at a time. Only by spending time with them will you be able to internalise how they sound and therefore be able to select and employ them with ease.

Checklist

- Do you know the five delay time categories?
- Are you familiar with tape, analogue, and digital delays and their subtypes?
- Have you explored and understood the differences between the modulated delay types (chorus, flanger, phaser)?
- Have you explored both tempo-synced and off-grid delay times?
- Have you reaffirmed the importance of having intention and being selective?
- Have you worked through all my top tips to ensure you understand how to process and apply delays in various contrasting contexts?

You should move on to the following chapter only once you can answer yes to all these questions.

Unit 7

Automation

Day 1 – What is automation?

Until this point in your mixing process, everything you have been doing has been static. This means that you've set something up, and it has remained that way indefinitely because nothing is altering it. This is a great starting point, but it doesn't make for an exciting mix. Whether noticeable or not, your listener will prefer a mix with some movement that doesn't remain the same from beginning to end.

This is where automation comes in. Automation is the process of gradually or suddenly changing the state of something during your mix. As we'll explore in this unit, what you can automate is pretty much limitless; your creativity is the only boundary.

This is the stage where your mix will start to come alive. And thankfully, in the modern digital age, it's simple to do. Within your DAW, you'll be able to automate any parameter of any plugin or channel easily. In the analogue domain, any automation move would physically have to be recorded or performed by the mix engineer. Whilst this could create some beautifully spontaneous tracks, it could also be very expensive. In the analogue world, every take costs money in terms of tape. However, in the digital domain, you can have as many do-overs as you like, safe in the knowledge that they cost you nothing.

In this unit, we'll explore different ways of recording automation, some great examples of which things are commonly automated, and how to maximise their impact without overdoing it.

Day 2 – Automation modes

When using automation within your DAW, you are likely to come across four different modes. Understanding these modes and differences is necessary to get the most out of your automation. How they're labelled in your DAW may vary slightly, but the concepts are the same.

The first mode is Read. This simply means that any automation that has been written onto your track will be read and therefore performed. The

DOI: 10.4324/9781003373049-7

automation won't be played back if this mode is not engaged. Always remember to change your mode back to Read once you have written in any new automation. This will ensure that you don't accidentally overwrite this information and that what you've written in will be performed and therefore heard.

The second mode is Write. This is also straightforward. Any parameter changes you make to any plugin whilst the track is playing will be written into the automation lane in Write mode. This is your first way of capturing a performance. Just remember to change the mode back to Read when you're done, and you'll be fine.

Mode number three is Touch. Touch mode is handy for making minor changes. It's helpful because when you stop moving any parameter, it will snap back to where it was before you began your alteration. So, you can make minor adjustments on the fly, making momentary tweaks that will revert to their original state when you let go.

The opposite of this is Latch mode. Latch means that the automation will latch onto the final position once you make your adjustments and remain there. This is great for creating more significant, long-term changes to your track.

The other major thing to point out is that most DAWs will allow you to write automation directly into a region or onto a track. This will affect how the automation is stored, processed, and accessed, so it's worth considering how you want to use it long-term before you write it in.

> TASK – Find how to turn automation modes on and off in your DAW. Identify what different automation modes you have available to you. Also, discover if your DAW will allow you to store automation by region as well as by track.

Day 3 – Automation types: Fades and curves

Having learned how to capture automation in your DAW, the next thing to identify is how automation can move. By this, I'm referring to the motion of the automation. The first and most obvious motion you will likely use is a fade. Automating a fade will allow you gradually to change the state of something over time. In this context, I don't just associate fading exclusively with volume but with the gradual change of any parameter. Therefore, fades are great for things like volume changes, alterations to effects send levels, sculpting your ADSR envelope in your synthesiser, EQ filters, and so on. Automating in this way will bring a level of subtlety to your mix. Having moving parts that slowly evolve will create a level of intrigue, encouraging your listeners to focus intently on the details of your mix.

But fades can provide more than just an even-paced slide from point A to point B. You can create an additional layer of sophistication and drama by adding curves to your fades. In music-production terms, we generally refer to fades as either exponential (getting faster and faster) or logarithmic (getting slower and slower). In automation terms, you can think of these in two ways: Exponential fades are great for ramping up the drama, creating hype and building tension. Implementing them leading up to your chorus or drop is a surefire way to generate the level of impact you're after. Logarithmic fades provide the converse effect: A winding-down, slowing feeling. Using these coming out of your chorus back into your verse or coming out of your breakdown is a solid place to start.

The condition for both curve types is that you pair them with automation going in the right direction to support the curve. For example, for an exponential volume fade to be most effective, it needs to be increasing in level rather than decreasing. In this way, the slower part of the curve will be where the level is low, and as the curve picks up momentum, the speed of the volume fade increases.

TASK – Practise drawing in automation fades. Try it on a range of parameters. Then see if you can add both exponential and logarithmic curves to these fades. Your DAW should have a function that allows you to add automation curves.

Day 4 – Automation types: Binary, step, and spike

There are three more automation motions that you can use. First, there's binary automation. In automation terms, binary refers to a hard change of state. Implementations for binary automation could be plugin bypassing, muting and unmuting channels or effects, turning EQ filters on and off, synth functions like turning oscillators on and off, transitional effects like glitches and bitcrushers, and so on. So binary automation is great for something you want to be instantly gratifying, providing immediate impact.

But sometimes binary won't cut it. If you want something more specific, you can use step automation. Step automation is a more particular version of binary. Here are some examples of uses: Synced LFO rates and oscillator pitch alterations, transposition, A/B crossover on synths or effects, rhythmic effects like gating effects, stutters, etc. This aggressive style of automation is not suited to all genres of music. It's best placed in faster-paced, more aggressive genres rather than smoother, more ambient music.

When a fade is too slow, binary is too limiting, and steps are too harsh, the solution is a spike. Spikes are effectively micro-fades or curves. Use them for transitional effects like filters, distortion and reverb, short filter automations like bringing up the cutoff quickly, and pitch alteration.

> TASK – Practise creating binary, step, and spike automations. Explore different parameters with each one. Are some things better suited to certain automation types?

Day 5 – Have intent!

I know I sound like a broken record, but it is such an important skill to develop. Like any other mix decision you make, you must have a reason for adding automation to your track. Analysing your mix critically, deciding whether it needs something else, and rationalising whether what you add is any good is a tough road to walk. It requires you to leave your ego at the door. If you're mixing for a client, as I often am, something that you think is perfect for the track may not sit well with them at all. Ultimately, the clients are footing the bills, so only their opinion matters.

In terms of your automation, ensure you've got a specific reason for any move you make. And if you find yourself unconsciously adding automation without giving it due consideration, stop!

Justifications for automation moves don't need to be complicated. You could want to add a high-pass fade during a buildup to increase the tension and help the drop to smack more. Or maybe you need to automate a reverb bypass at the end of a section to signal a new musical idea. You might find that a particular synth is getting boring, so you decide to write a few spikes into the filter cutoff to maintain interest.

By validating each automation move before, or at least whilst, making it, you guarantee you're not throwing things at your track that will end up accumulating into a confusing mess.

> TASK – Review a previous mix. If it has automation present, sense-check each move. Do they all have a purpose? If there is no automation, consider if it may benefit from some subtle movement.

Day 6 – Bouncing automation

If you're an amateur producer working on a consumer-grade computer, it's at this stage in your mix process that you are likely to begin to stretch the capabilities of your machine. From experience, there's nothing more frustrating than showing a mix to a client only to hit a system overload halfway through the track.

Automation is particularly CPU-hungry. It's obvious when you think about it: A static track with no automation is easy to compute. Your machine will look at it once when you hit play, realise it's staying the same all the way through, and therefore relax. However, when your machine sees a track with automation present, it knows that it needs to watch that track the whole way through in case anything changes at any point. In other words, a track with automation needs to be constantly monitored. The more channels you have with live automation, the more likely you will encounter the dreaded system overload.

A good remedy here is to bounce your tracks out. Different DAWs facilitate this in different ways. Freezing or rendering amounts to the same thing, too; they're variations on the theme. The priority here is to commit an automation move into the audio region so that it doesn't need to be processed in real time. Your focus here should be to commit to any software instrument parameter changes you're making. Running any software instrument, especially a sample-based one, will be demanding on your CPU – to have moving parts within that instrument even more so. It's good practice to mix with 100% audio and 0% MIDI anyway.

Now you're probably thinking, 'That's all well and good, but what if I need to go back and change a part or tweak a sound?' That's not an issue. Once you've bounced the track out to audio (committing to the automation moves in the process), go back to the MIDI track, turn off the software instrument and any other plugins on the channel, freeze it, mute it, and hide it. It'll all still be there if you need to go back and make changes, but it won't be fogging up your system in the meantime.

> TASK – Practise the process of bouncing MIDI channels to audio and then turning off and hiding the original MIDI track. Remember, the objective is to free up the CPU without losing any information you may need to amend later.

Day 7 – Three stages: Problem-solve

To help you make better decisions when writing in automation, you can ultimately think of automation in three stages to help speed up your workflow. The first of these is problem-solving.

This stage is all about finding apparent problems with your mix and fixing them. A problem doesn't mean that there's something wrong, per se. Anything truly wrong should have been repaired well before now. It means that something is detracting from the overall success of your mix. Some examples of problems that need solving could be that your intro has too much energy or that your lead may be too dry in one section but too wet in another. You could solve the energetic opening by automating a high-pass filter on your kick to reduce the low-end content. You could fix your lead by automating the dry/wet blend or send amount.

This stage is all about solving obvious issues; things that are detracting from the overall success of your mix. Don't get hung up on looking for creative nuggets of ear candy at this stage. That will come later. Focus on identifying things that noticeably detract from your mix. If it's catching your ear negatively, then it's likely to be catching a listener's ear in the same way.

> TASK – Review a previous mix. Can you identify any problems that could be addressed with automation? Fix them!

Day 8 – Three stages: Flow

The second stage of automation is all to do with the flow of your track. This stage is about ensuring that your song moves seamlessly from one section to the next without anything sounding jarring or abrupt. In a way, this could be considered as a second problem-solving stage. But here, you're not looking for things that are jumping out at you. Instead, you're looking for slightly rough edges or minor imperfections that you can smooth out. If the problem-solving stage is like ensuring that the piece of furniture you're making is built absolutely to the plan, then the flow stage is like sanding down any saw marks that may have been left in the process.

You should have a particular focus on transitions at this point. You will be likely to use many fades and curves with a keen emphasis on automating volume levels, effects levels, and EQ filters. Your attention to detail will make all the difference in the quality of the outcome. You may not think that a 1dB cut here or a 0.5dB boost there will make any difference, but it will! The success of your mix will primarily be measured not by what your listener notices but by what they don't. By this, I mean that if your listener isn't distracted at any point by any mix elements, they will therefore be able to engage in enjoying the track fully. Arguably then, this stage is what separates the good from the great. The ability to identify minuscule elements that require the smallest amount of polish takes time and energy to develop.

> TASK – Review the transitions of a previous mix. Are there alterations you can make to the differential between sections that will enhance the flow of the track?

Day 9 – Three stages: Create

Having solved all your problems and sanded out any imperfections, it's time for the fun part: The creative stage. This can be considered as putting the icing on the cake. However, proceed with caution. Whilst icing is most people's favourite part of a cake, too much of it will throw out the balance, making the whole thing too sickly sweet. Whilst there are no limitations, you want to avoid detracting from the song at all costs. Yes, this is your chance to show your competency as a producer, but you can demonstrate competency through restraint. You should look to add just enough flair and complexity to make your mix shine without making people's stomachs turn.

One way to ensure you don't overdo creative automation is to consider it subtractively rather than additively. This means that when looking to creatively draw the listener's ear for a moment, you do it not by adding something that steals the focus but by taking something away. By subtracting something from your mix, perhaps by automating the level or send amount or filtering aggressively, you'll find that you create space in your mix for something else to shine through. So, rather than forcing the attention onto something by adding in level, you can deflect attention by creating space. To me, this is a more intelligent, subtle, and overall, more experienced way of creatively automating that demonstrates knowledge and experience.

TASK – Find something in a mix you want to draw attention to. Rather than enhancing it by raising its level, find a way to create space for it by reducing other things. Don't be limited just to level. Think about EQ and pan position too.

Day 10 – When should you automate?

You now know all the different automation types that are available to you. The big question on your mind is, 'When exactly in the production process should I be automating? After arranging? After mixing?' Whilst there isn't a hard, fast rule, my advice is simple: Do all your automation after setting up your static mix.

Let's expand on this point: One of the biggest mistakes I see amateur producers making is being unable to distinguish the boundaries between the different stages in their workflow. The writing stage gets blurred into the arrangement stage, which gets smudged into the mix stage. Not having boundaries between your different processes makes it impossible to decipher when you are moving on from one process to the next. Therefore, you're never truly sure if you've finished writing your song or have finished arranging it because your mix process is caught up amongst everything else.

By learning to define when one stage in your process ends clearly, and the next begins, you'll be able to progress more confidently and, therefore, more efficiently.

Knowing that your automation comes after you have set up your static mix takes it off the table until that point, keeping your brain focused on the task at hand. Allowing your mind to focus on a single job at a time will enable you to execute that job to the best of your ability as you won't have split foci – you will create tunnel vision for yourself.

The single exception to this rule is if you need to create automation for sound-design purposes. If you need to automate a particular synth parameter or filter to help you build a clearer picture of how your track will sound, then, by all means, do this during your arrangement stage.

The clear justification for saving automation until post-mixdown is this: It needs its final context to work in to know whether an automation move is working in your track or not. There's no point automating a level change or effects-send change until you see what it will be competing with at that moment. Otherwise, you'll be guessing. Even things like automating cutoff frequencies and filters need context. Whilst you may want to lay down markers in your project to note where you want certain things to move later, I would recommend keeping a list of automation moves you know you want to make in your project notes so that they can be dialled in with accuracy at the right and proper time.

> TASK – Less of a daily task here and more of a shift in approach. In forthcoming projects, leave your automation until your static mix is complete. It will feel challenging to begin with, but after you've done a few, you'll find your static mixes slot together more efficiently, and your automation makes more sense in context too.

Day 11 – Use No. 1: Levels

Now that all our ground rules have been established, we can begin discussing some real automation applications. Let's start in the most obvious place: Levels.

Automating the overall level of something in the mix is the most basic yet most crucial automating skill to nail. Yet, it has more depth to it than you may at first think. Consider a drum kit that has been recorded with an extensive range of microphones. Your goal is to create a contrast between the verses and choruses of your song. The first thing you decide to do is to bring the level of the kick-out mike (the mike outside the resonant head of the drum) down in the verses to focus more on the kick's attack from the kick-in mike inside the drum. When you bring back up the kick-out in the choruses, it will add nice weight. Secondly, you decide to bring up your stereo room mikes in the verse to provide more contextual space to the kit whilst the arrangement is sparser, compensating by slightly bringing down the level of the spot mikes. In the chorus, you reverse these moves to get more of the direct sound and decide to increase the overall amount of parallel compression in your drum buss to make your dynamic more consistent. These small moves could make all the difference to the overall texture of your track.

You can apply this principle to anything: Perhaps multiple mikes on a guitar amp, a clean and distorted bass channel, or the amount of double-tracked vocal sitting underneath your lead.

But this is only 50% of the conversation at this point. While considering the overall level of musical sounds, you should also consider your effects. Treating effects as another instrumental component will help you truly embed them into the mix. You may wish to bring up the level of your vocal reverb or delay in the verse to make it feel more reflective but reduce it in the chorus to get that strong, direct sound. You may have some flanger or phaser on your guitar but decide that it's more appropriate just to have it featured in the bridge of your track.

You can create so much interest in your mix just by automating the levels of instruments, vocals, and effects. It's worth spending a reasonable amount of time experimenting just with level automation to explore its capabilities thoroughly.

TASK – Focus on automating levels of things to create subtle depth and variation in your mixes. Ensure you include effects in your thinking.

Day 12 – Use No. 2: Stereo width

I alluded to this earlier in the course, and it is now time to discuss it in full. Automating the stereo width of something or automating something's position to create more or less width is a brilliant weapon to have in your arsenal. This device is handy when trying to combat the age-old problem of making a point of difference between different sections in your song without doing anything overly dramatic. Let's look at some examples.

Consider your drums again. Depending on the sort of drums you're working on, i.e., electronic or acoustic, and how they were recorded, i.e., a few mikes or lots of mikes, you will have options available to you. You could try using the mono room mike in the verse and swapping it for your stereo pair in the chorus. If you don't have a mono room mike, you can use a directional mixer on your stereo pair to make them mono. If this effect is too much, you could simply narrow your stereo mikes in your verses and widen them out in your choruses.

Consider pads as another example. Typically, your pads will be a stereo sound, utilising the entire width of your stereo field. But what if you narrowed them in the verses and widened them in the choruses?

What about guitars? If you have double-tracked rhythm guitars panned hard left and right for your choruses, you could narrow their position in the verses.

What could you do with your vocals? You could use a short stereo delay to create width in your choruses on a single lead vocal. If you have a double and triple-track, you could use just the double-track to support the lead in the verse but use both in the chorus, moving them left and right of the lead to create width. You could use mono backing vocals in your verses placed in specific locations in your stereo field but use stereo backing vocals panned hard left and right in pairs in your choruses.

There is so much that you can do to create contrast in your stereo image. However, the caveat is always not to get carried away. With so many options available, you can effortlessly throw automation at almost anything that crosses paths with your cursor. Allow yourself to slow down, ensure you ask yourself the critical question: 'What am I trying to achieve here?' and you should be fine.

> TASK – Focus on automating the width and stereo position of things to create subtle variations in your mixes.

Day 13 – Use No. 3: More automation, less compression

A problem producers encounter repeatedly is getting their vocals and bass to sit nicely in their mixes. People often complain of vocals being on top rather than within their mix, bass feeling uneven and not grounding their track, vocals getting lost, and so on. The usual fix that people turn to is to compress the life out of everything to try and create an even signal that is easier to balance. But inevitably, this leads to a lifeless track that has no depth or musicality left in it.

The resolution that I always suggest is to automate the signal rather than compress it to the point of consistency. By automating a track, you can specify precisely how loud or soft you want that track to be at any point in time. This gives you a greater level of control than any compressor could.

On the face of it, this sounds like quite a long-handed solution to a simple problem. Going through a song and automating the level of your bass and vocals is a lot of effort. In the analogue realm, you'd physically record your fader automation to tape in real time.

The good news is that there are plugins out there in the digital domain explicitly designed to tackle this problem. For example, Vocal Rider and Bass Rider from Waves ride your faders automatically. You simply set a target output level as a threshold, and away you go.

The outcome of automating your level rather than compressing it is that you end up with a signal as upfront as you want at any given moment whilst retaining the musicality and dynamic range you'll undoubtedly want to preserve.

> TASK – Practise automating vocal and bass parts to maintain a consistent level throughout your track whilst avoiding over-compressing.

Day 14 – Use No. 4: Fix plosives and sibilance

Another common problem home producers face is the issue of plosives and sibilance. Many factors influence the recording environment when recording vocals at home, potentially leaving you with a less-than-desirable result. The microphone you have available, the space you are recording in, and the ambient noise in that space will all impact the quality of your recording. It's common to end up with an undesirable amount of plosives, sibilance, or both, typically caused by the singer being too close to the mike.

Sibilance is the easier of the two to address in your mix. The job of de-essers is, primarily, to deal with excessive sibilance (although they can be used on any audio source, e.g., on hi-hats and cymbals, to tame harsh upper frequencies). A de-esser is effectively just a dynamic EQ. You set your target frequency and allow the compressor to duck the targeted region. However, when used with a heavy hand, de-essers end up sounding as if the vocalist has a lisp, so you need to be cautious!

Plosives are more challenging to deal with. Once a plosive is baked into the take, there's not a great deal that you can do to fix it. Yes, there are expensive audio repair plugins that can help, but those sorts of tools certainly won't be in every amateur producer's toolbox.

My suggested remedy here is, again, to automate. It seems like such a simple fix, but it's the most effective. When you have a plosive jumping out from the performance or an overtly sibilant syllable, simply duck the level with automation. In this way, you don't risk contorting the vocalist's natural voice through heavy-handed de-essing. You will need to do this manually, I'm afraid. But in this context, that's for the best. You're likely only to need to duck a few syllables in your whole song, so it's worth being meticulous here.

> TASK – Automate a vocal track to duck some excessive plosives and/or sibilance. Compare this with a de-esser. Does one do the job better than the other?

Day 15 – Use No. 5: Increase plugin control

The quickest, most straightforward way to arrive at a mix you can be proud of is to maintain control over it from start to finish. This is why I so heavily laboured the point about gain staging earlier in this course. By staying in control throughout, you ensure that you know precisely what is going on in your mix at any given time. The result is that if you hear something unpleasant, you know exactly where to look to address it. You want to avoid rummaging around in the undergrowth of your mix, looking for an issue you can't quite put your finger on.

Automation is a big help in this regard. Being able to automate any parameter of any plugin means that you can stay 100% in control. Let's look at a few examples to quantify why this is helpful.

Consider a plugin you are using as an effect to create drama at a specific moment. You only want that plugin to be on for a short time. By automating the plugin's bypass, you ensure that the effect only operates when you tell it to, preventing any potential oddities.

Consider a delay effect. A common technique is to automate the delay time to draw attention to the effect at specific moments. Automating the time ensures that your wet/dry blend, which you presumably dialled in meticulously, remains the same. Therefore, you retain control.

Consider a distortion or saturation plugin. You may well want to push the distortion amount, perhaps in the chorus or during a solo. By automating the distortion amount, you are guaranteed to receive your desired amount.

There are so many examples of how automation allows you to maintain control over your mix. What about pitch shifter effects, filter resonances, and cutoffs?

> TASK – Today's task is a reflection. Think about any time you could hear an issue in your mix but couldn't locate it. Think about how frustrating that was. We've all been there! Now, use that frustration as the catalyst to stay in control of your mixes moving forward.

Day 16 – Use No. 6: Aux sends

This point has been touched upon already, but it's well worth labouring. Auxiliary sends shouldn't be set up once during your static mix and forgotten about. The amount of effect you will need in your mix will be linked to the context within which it appears. I mean that the amount of reverb, for example, that you need on your lead vocal will be different in your chorus to your verse. This is because the bed in which it lies is different. There will be changes in instrumentation, texture, dynamics, etc., all of which will affect how your vocal comes across.

You can automate any auxiliary send amount. I thought it would be helpful to focus your mind on the most common aux sends you're likely to need to automate.

First is reverb. For precisely the reasons I have just mentioned, you should be automating your reverb send amounts. Start on the most prominent aspects of your mix. Your reverb send level will be far more audible on your lead vocal than on a background pad. Once you've worked through the most focal points in your mix, you may find that's enough. The need for automation on effects sends becomes less the further back elements are in your mix.

The same principle should be applied to your delay sends, but with a slight twist: Your delay amount is more likely to need to be looked at on a phrase-by-phrase, or even word-by-word, basis. This is because, depending on the length of the phrase you're working with, the space you'll have in between for the delay to shine through will differ. Therefore, you're more likely to want to automate your delay on a smaller scale.

The other two leading players here are parallel compression and parallel saturation. Saturation is a topic I'll talk more about later in the course, so don't worry if you're currently totally baffled by the term. In simple terms, it means distortion. Contrary to popular belief, not all distortion is bad. A bit of distortion is an excellent thing. To put things as simply as possible, when you saturate something, you're adding additional frequency content, which will fill out and round out the sound. So, this is an excellent device for making something more prominent at a particular point in your track.

Parallel compression works in a not-too-dissimilar fashion. It's a brilliant tool for helping parts stand out in a mix without removing the dynamic range and musicality from the performance. This is because you're blending a heavily compressed signal back in underneath the natural performance that still has all the undulations and fluctuations in it. Pushing your parallel compression level on, for example, your lead vocal during the chorus, or your lead guitar in the guitar solo, can help it cut. But you certainly wouldn't want that amount of parallel compression present throughout.

TASK – Review a previous mix. Have you automated any aux send levels? If not, use this opportunity to do so.

Day 17 – Animate your transitions

One of the most important places to focus on when considering automation is your transitions. How you move from one section to the next is crucial to your mix's momentum and overall drama. If you simply flop or stumble from your verse into your chorus or from your chorus back out into your verse, you're likely to be doing yourself a disservice. However, get your transitions just right, and you'll maintain all the drive and energy you're after.

We've already looked at ways to add contrast between sections, automating stereo width, filters, parallel processing, etc. But one of the simplest ways to add real impetus to your transitions is with a swell. And there are several different types that you can use.

First, there is the simple volume swell. By automating the overall level of something from low to high, you'll add weight to the impending section. This is a common technique on electric guitar, where volume pots are quickly and easily turned from 0 to 10.

Secondly, there is low-pass filter automation. Automating your filter so that it sweeps from somewhere low around 500Hz all the way up to the high end will gradually introduce a cool swell.

Thirdly, and my personal favourite, there is the reverse cymbal. You create this by grabbing a cymbal hit in isolation (if you're using one-shots, so much the better) and reversing the audio region. If you combine this reversed sample with a volume swell, you'll create real drama.

Lastly, you can use a riser or uplifter. These terms mean the same thing. You can find thousands of premade samples, some that are pitched to key, some that aren't, of varying lengths. In general, they combine all the above. They'll be some sort of volume-swelling, filter-sweeping, reverse-cymbal-esque effect!

Use any of these to grab attention and up your transition game.

> TASK – Experiment with the transition effects outlined previously. Ensure they are texturally complimenting your transition rather than overpowering it.

Day 18 – Automate synth parameter changes

Another great place to add further interest to your mix is to automate parameters within your synth instruments. This is where we hit the grey area that I mentioned earlier. You may well cover this stage within your writing or arranging process. Wherever you do it, though, here are some of the main places to look to add interest. As a quick disclaimer, if I use terms you don't understand here, don't fret. I'll be covering the fundamentals of synthesis later in this course.

First, probably the most common parameter to automate is the cutoff. In most synths, the type of cutoff frequency is changeable, so you can make it a low-pass, high-pass, or even band-pass filter. Therefore, it's excellent for making a point of difference between sections and for transitional effects.

Next is your oscillator blend. Generally, synths will come with at least two oscillators. Each oscillator can create an entirely individual sound. The oscillator blend allows you to mix different amounts of the two. So, you could have more A and less B in the verse and vice versa in the chorus, for example.

Thirdly, I'd look for the attack and release time in your envelopes. Particularly regarding the amp envelope, adjusting the attack and release times will allow you to create more or less clarity at the front and back end of your synth lines. This will allow you to push parts backwards and forwards in your mix without needing to reach for things like reverb or volume automation.

Another great thing to automate is the resonance. This puts a frequency bump just before your cutoff frequency. Adding a resonance boost will help your synth to stick out. This is useful if you want to pull your synth forward for a few moments before pushing it back into your mix.

I also like to automate the pitch of my oscillators, particularly the cents value. Increasing or decreasing the cents of one oscillator will create more of a chorusing effect with the other. This is great for thickening your sound. Just don't go so far that you make it sound out of tune (unless that's what you're after).

There are loads of other creative ways to automate your synths. These are just a few to get you going.

> TASK – Find a track in which you've used a synthesiser. Find the parameters mentioned above and experiment with them to create contrasting sounds. Automate these tastefully in a mix.

Day 19 – Emphasise drum fills

Let's consider drum fills in our automation conversation. To understand drum fills, you must first appreciate the instruments within a drum kit. On a standard drum kit, you have five drums: Your kick and snare (generally the backbone of your track) plus three tom-toms (high, mid, and low in pitch).

You then have cymbals: Hi-hats and a ride (used to keep time in the groove, typically with eighth or sixteenth notes), and one or two crashes (to accent important beats). A drummer has four limbs, so they can play a maximum of four of these eight or nine components simultaneously. So, if a drummer can play half of the kit simultaneously, how do you create a point of difference for a drum fill to make it stand out?

The obvious place you must start is the arrangement, of course. Carefully arranging what instruments are used with what rhythms at specific times will add importance to your fills. Then there's the performance, which will make or break how the fill projects in the mix. But once you get the performance into a densely populated mix with other things going on around it, you may find that your fills lack weight.

The simplest way to solve this is to automate your drum fills to add the required amount of weight. As with most things, you have options available to you here. You can reach straight for the overall level. This will work fine. You could reach for your parallel compression amount. This will be slightly more subtle but perhaps more tasteful and less noticeable. If you're working with acoustic drums, you could be a bit more creative and bring out your fills by increasing your room mikes. You could even increase your saturation level.

As you've probably worked out by now, there's more than one way to skin a cat, as the expression goes. The takeaway point is that automation is your friend when adding weight and importance to your drum fills.

> TASK – Experiment with some different methods of adding weight to drum fills. Do you have a preference?

Day 20 – Tidy up

I absolutely love using chokes in my writing. A choke is where all musical content stops simultaneously. Whether it's a choke before a heavy drop or a choke to end a track, I think both sound great. I'm probably guilty of using them too frequently. But all too often, I hear them used poorly in amateur productions. There are two possible ways to do a choke badly.

The first way is not to do anything at all. This means that whilst you've written a choke into the arrangement with all instruments performing it, you haven't produced it into the mix. This will inevitably leave you with multiple tails ringing through what you want to be a choked space. You should be looking out for cymbal decays, guitar tails, synth release times, etc. – any potential sound that could be ringing. You can tidy up these things with volume automation. You should also look for reverb and decay tails that ring through.

Generally, rather than choking the level of these time-based effects to zero, you'll want them to duck but still be audible to keep a natural feel.

This leads nicely onto the second way people get it wrong: They overproduce a choke. You can do this by reducing the level of time-based effects to zero, effectively creating a complete silence in the track, which will stick out like a sore thumb. But you can also get it wrong by automating your instrument levels too early. Typically, this happens when trying to produce a choke into a poorly performed part. This results in cutting off parts before their natural break. You want to look for the moment just after the drummer grabs the cymbals to stop them ringing, just after the guitarist mutes the strings, and so on. Look for the natural break in the track. A tight performance is critical to ensure you can make the most of the choke in the mix.

> TASK – Tidy up the chokes in your track. If you haven't got a project that contains a choke, write one into something new!

Day 21 – Accentuate vocals

Accentuating a vocal throw is one of my favourite ways to emphasise a specific lyric or phrase. But technically, it's not actually an automation process. I've included it in this chapter because it sounds as if it should be!

A vocal throw is usually applied to a lead vocal, but the method can be applied to anything you want to add importance to in your mix. The concept is to select a small idea (this could be a short phrase or even a single word or syllable) and throw it to the left and then to the right of your mix. You achieve this by copying the audio you wish to throw, duplicating it onto two new tracks, and panning them as hard to the left and right as you desire. You then move the throws backwards, so they come after the main sound. So, you're effectively manually creating a ping-pong delay. It's normal to offset one by a quarter note and the other by a half note. Then you'll probably want to filter the throws so that they are pushed further back in the mix and wash them out with some reverb.

Of course, you can achieve all these steps with a decent stereo delay plugin (except for the reverb). But that's quite a few automation steps to go through. I think it's easier to maintain control over the throws when they're on their own audio tracks. It just keeps things more manageable.

Throws generally work best in sections of a track that aren't too busy where there is space for them to carry. If you try to employ one in a densely populated section, you'll find it's tricky to get it to cut through the mix.

Of course, you can apply this principle to anything. You could try it on an individual snare hit, a guitar bend or slide, or even a horn or synth stab. Get creative with it!

> TASK – Manually create some throws. You can try it on a vocal or any other instrumental part to which you wish to add importance.

Day 22 – Automate panning for sound design

Sound design is something that has become more and more prevalent in modern music-production. But to understand it, I should define what I mean by it first. In film and visual media, sound design is how the aural world is fleshed out to enhance atmosphere, tone, and mood. In the world of music, it means the same thing, with added emphasis on enhancing the song's drama.

The most common use for sound design is to enhance transitions. We mentioned earlier uplifters and risers. These go hand in hand with impacts, downshifters, and, to a lesser extent, sub-drops. You can think of an impact as a short sound used to punctuate the beginning of a new section with emphasis. A downshifter is the reverse of an uplifter, so it has the effect of releasing tension. A sub-drop is essentially an impact but solely focused on the sub-bass frequencies to provide a gut punch when listened to on a system large enough to reproduce these extreme low-end frequencies.

The other everyday use of sound design in music is to provide contextual or background noise or atmosphere. For example, it's commonplace in dance music to add tape hiss or vinyl crackle into the audio bed. In this category, you can include anything that will be layered subtly in the undergrowth of your track to subliminally reinforce a feeling, mood, or atmosphere without taking focus from any other element in the mix.

In all these cases, the sound design is likely to benefit from some level of automation. It could be that its level needs automating from one section to the next. Perhaps you want to pan it from right to left and back again. Maybe you want to automate its reverb send or its low-pass filter. There are infinite ways to get creative here.

> TASK – Implement some different sound-design options into your track. Assess whether they would benefit from some automation to bed them in more naturally.

Day 23 – Automate EQ

As you know already, EQ is one of the Big Four when it comes to mixing. And I've already mentioned it in the context of automation with my references to filtering. But there's much more that you can do with it. Let's look at some examples.

Think about your classic low-end conundrum. You feel you've balanced your kick and bass well, and they're working throughout 90% of your track. But there's this one moment where you want the low end to feel even thicker. The best way to achieve this is by automating an EQ band for that specific point in your track.

Consider a lead guitar or lead synth part. For the most part, it's sitting well within your mix, complimenting your vocal but not obstructing it. But there are short windows of opportunity for it to shine through a little more. Rather than automating the volume, which could feel too obvious, you could automate a presence boost in the lead line to bring it forward a bit for those specific moments in time.

Or, you've got a lovely breakdown in your track where you want to hear the lead vocal reverb shimmer a bit more. Rather than automating the whole send amount, you decide to boost the upper frequencies of your reverb with a shelf.

Having read these examples, you may be thinking, 'Can't I achieve all of that through dynamic side-chained EQ?' The answer is yes, you could. You've already realised that there is often more than one way to achieve the desired result in the world of music-production. How you choose to get from point A to point B is a personal decision. My job here is simply to make you aware of as many potential routes as possible. The journey is then yours to take.

> TASK – Practise automating EQ to make things come forward at certain moments in a track.

Day 24 – Make e-drums sound more natural

One of the easiest ways to spot an amateur mix from 100 yards away is by listening to the drum production. Experienced producers know that real drummers play with expression and that it's almost impossible to make one drum hit sound precisely the same as the one that came before it. Therefore, they spend time editing their MIDI to ensure their drum lines are packed with 'human' traits. Amateur producers haven't perfected this yet, and therefore the drums sound mechanical and stiff. There are a few obvious places to look to improve your drum programming dramatically.

First, look at your velocities. A human drummer will not strike a drum with the same velocity twice in a row. However proficient they are, it's almost impossible to nail the same dynamic repeatedly. With e-drums, you can adjust your velocity of notes to recreate this. The long-handed way to do this is to change each note individually in your piano roll or step sequencer. Doing so will give you maximum control. The short-handed method here is to automate the velocity in your sampler. If you automate your velocity with an LFO that oscillates against the timing of your track, you'll get some very useable velocity variation.

The second place to look is at the pitch of the sample. Each time you hit a real drum, the pitch will be very slightly different depending upon the position on the drum and the velocity with which you strike it. By automating your sample's pitch, you can, to a certain extent, replicate this effect. You'll want to affect the fine rather than the coarse pitch, meaning that you're affecting cents rather than semitones. Again, an LFO here means you can set it up and forget about it, rather than having to write in constant streams of automation.

The last, less obvious place to look at is your sample's release time. The decay on a drum will often vary depending upon how you tune the batter and resonant heads, where and how hard you strike it, the stick type, and so on. Subtly automating the amp envelope's release time will give you some of this gentle variation that will take your production away from the amateur and towards the experienced.

Of course, all these principles apply to cymbals as well, most importantly hi-hats and rides. Because the frequency with which you strike these timekeeping cymbals is so much higher than your drums, you can be more pronounced with your velocity, pitch, and release variations.

> TASK – Practise programming electronic drums with a focus on a human performance. Look to automate the parameters mentioned previously to keep things natural.

Day 25 – Tempo-synced automation effects

A great place to use tempo-synced automation is with auto-pan effects. For example, if you want a sound to pan from left to right in time with the music, drawing in your automation so that it's locked to the grid will ensure that the panning stays in time. In most DAWs, you can draw automation using different waveshapes. Experimenting with different waveshapes here (sine, triangle, saw, square) can be fun. Each shape has its own feel. However, it's easy these days to set up these sorts of effects in a designated auto-pan plugin. So, what else can you automate in a tempo-locked fashion?

Try automating effects levels. Consider how different a delay send or a distortion send will sound when tempo-synced with a square wave versus a sawtooth wave. The square wave will act as an on-off switch, whilst the sawtooth will gradually slow down with a quick switch back on.

You can experiment with track levels in this same tempo-synced way too. The waveform that you choose will denote the overall feel. For example, a backing vocal coming in and out on a sine wave will feel smooth, whereas a pad that chops in and out on a square wave will feel more aggressive.

A creative way you can use this effect is on filters. Think about the sound you'll achieve by tempo-syncing the automation on a low-pass filter.

If you like the direction in which the effect is taking you, but it feels a bit too much, remember you don't need to go hard on it. If you're auto-panning, you don't need to go from hard left to right. You could go just 50% either way. If you're automating track levels, you don't need to go all the way off and on. You can reduce by any amount that you feel compliments your mix well. Small details always add up, and three or four small changes will probably sound better than one massive one.

> TASK – Experiment with some tempo-synced automations. You can use the previous examples to get you started.

Day 26 – Automate your master buss

I've come across multiple 'sources' that tell producers not to touch their master fader. They state that your master buss should always remain at unity and that you shouldn't do anything else with it. Whilst there is a sliver of truth in this, it's not a very useful statement. Let me explain why.

Your master buss, mix buss, two buss, stereo buss, or whatever else you want to call it is the most critical part of your mix. It's where you try to tie all the elements in your mix together to make them sound as if they belong together as one cohesive recording. So, it makes sense that you don't want to do anything drastic with it, as this will skew your whole mix. This statement is the sliver of truth to which I previously referred.

But, for the same reason, it's also your only opportunity to do things with your whole mix. And that is undoubtedly an opportunity too good to pass up. Let me take you through some simple things you can do with your master buss.

First, a bit of gentle fader automation is a good thing. This is your opportunity to apply some subtle dynamic changes to your whole mix. And when I say gentle, I mean 0.5 of a dB here and 1dB there. Not great big moves that will be clearly audible. It should go without saying that you don't want to exceed unity with your master fader at any point. You may be surprised by the difference a 1dB boost into your chorus may have in giving it that additional weight you've been looking for.

In the same vein, you could gently automate your mix's stereo width. Narrowing your verses by 5% can make your choruses open out even more.

What about applying a high-pass filter to your verses? A high-pass filter at 35Hz that rolls off in your chorus to let that extra bit of low end through can make all the difference to your drop.

The first caveat with all these things is that I advise you to do them once you've finished the rest of your mix. It would be counterproductive to automate your entire mix before having finalised individual components. The second is that less is more here. You shouldn't be looking to make changes on your master buss that are clearly audible in your mix. They should be subtle changes that are felt rather than obvious moves that are noticeable.

> TASK – Implement some subtle mix buss automation. Try volume and stereo width automation as a starting point.

Day 27 – Automate your tempo

My final piece of advice in this chapter is to automate your tempo. I appreciate that for some genres, this won't be applicable. For example, you'd be foolish to have tempo changes in a techno track. It would make it impossible for a DJ to mix.

So, perhaps it would be better to say, 'Don't forget that you can automate your track's tempo'. Why is this an important consideration? To understand this, you need to consider where popular music originated.

Allow me to give you a very brief, extremely general history lesson. If you consider some of the earliest names in the world of popular music in the '50s and '60s, you think of the likes of Bill Haley, Elvis Presley, The Beatles, and The Beach Boys. At that time, recording to a click track simply wasn't a thing. A great deal of the expression of their recordings came from the subtle ebbs and flows in tempo. This was no new thing. The technique known as recitative was coined formally during the Romantic period and was one component in the desire to break away from the rigid rules of the Classical period.

It was only when synthesised and electronically generated music became prevalent in the '80s that tempo-syncing became a thing. And whilst it was stylistically appropriate for dance and other forms of electronic music, the application of a click track eventually integrated itself into almost all forms of popular music.

Nowadays, it's standard practice to record with a click. You'll go into a studio, and the engineer will say, 'What's the BPM, mate?' Professional drummers are well used to being plugged into a metronome, even on the live circuit. This allows for all prerecorded elements and even lighting cues to be synchronised to the millisecond.

Whilst there are some obvious benefits to this, especially concerning large-scale productions, it can have the added effect of making the music stiff and lifeless. The tempo of a track is, in my eyes, like the beating heart. It should be regular but not predictable. There should be some subtle variation that you feel rather than notice.

The beauty of working in a modern DAW is that you can write your tempo automation into your click track. So, you can give your track a 2 or 3 BPM increase into your bridge if you so wish.

The singular, crucial consideration here is that you should plan this information out before you start recording. This means that you must have written your track first to know your structure and, therefore, where you wish to input tempo automation. Then, once you've written this tempo map, you can record your song to it. The alternative in some DAWs is to use their 'smart tempo' functionality. This means you can record one part naturally (without a click) and then allow your DAW to analyse it and create a tempo map based on it.

Writing a tempo map before recording is only relevant if you are working with audio. If your track is purely constructed from MIDI, you can adjust the tempo wherever you like without issue.

> TASK – If you have a song that may benefit from a new, more expressive tempo map, use that. If not, create something new in order to explore this topic.

Day 28 – Unit summary

Hopefully, during this unit, you've realised just how crucial automation is in bringing your mix to life. This is where you take your track from the static to the dynamic, from the good to the great. As with most things we've discussed thus far, the critical message to take away is that less is more. I'm sure you've heard the expression that something is greater than the sum of its parts, and that is certainly the case here. An accumulation of subtle moves adds up to something great, where one large move just won't cut it.

Let's do a quick recap of everything we've discussed here:

- What automation is
- Automation modes: Read, write, touch, and latch
- Automation types: Fades, curves, binary, step, and spike
- Bouncing automation
- The three automation stages: Problem-solving, flow, and creative

- When to use automation
- Different uses: Automating levels, stereo width, more automation for less compression, fixing plosives and sibilance, increasing plugin control, and aux sends
- Animating transitions
- Automating synth parameter changes
- Emphasising drum fills
- Cleaning up chokes and endings
- Accentuating vocals with throws
- Automating panning for sound design
- Automating EQs
- Making e-drums sound more natural
- Tempo-synced automation effects
- Automating your master buss
- Automating your tempo

If you weren't sure just how deep the automation rabbit hole was before, you certainly are now!

By this stage in the course, having had seven months to practice and implement many of the lessons you've learned so far, you should be seeing considerable changes in the overall quality of your mixes. Until now, I've focused on working with the musical information that is within your session. I'm now going to change course.

In the coming units, we'll look at all things vocals and synthesis. I could easily have covered these topics at the beginning of the course, but I felt it was better to cover the mixing process in full first before distracting you with capturing and creating sounds.

So, without further ado, let's talk about vocals.

Checklist

- Do you understand all your DAW's automation modes?
- Can you write automation into regions and tracks?
- Can you create exponential and logarithmic automation curves?
- Can you create binary, step, and spike automation?
- Are you familiar with bouncing MIDI to audio and the benefits of doing so?
- Are you familiar with the three stages of automation? (Problem-solve → Flow → Create)
- Have you adjusted to automated after your static mix is finalised?
- Have you explored the subtleties of each of the six uses of automation?
- Have you explored transitional effects?
- Have you explored automating synth parameters?
- Have you automated drum fills?
- Have you automated chokes?

- Have you explored throws?
- Have you experimented with sound design?
- Can you make an e-drum part feel human?
- Can you create tempo-synced automations?
- Have you explored mix buss automation?
- Can you automate a project's tempo?

Unit 8

Vocals

Day 1 – Vocals 101

Probably the most frequent question I am asked on my social channels is, 'What's in your vocal chain?' Let's clarify; there is no such thing as the 'ideal' vocal chain. There are so many parameters that affect how vocals sound (as we'll explore in this unit) that you should treat every vocal recording as an individual. Even the same singer through the same mike in the same room in the same position may sound different from day to day depending upon how much sleep they got, what they ate for breakfast, their mood, and many other factors. So, if you were previously searching for a template vocal chain, hit the reset button, discard that POV, and prepare to retrain your mind to treat every vocal with the respect and individualism it deserves.

The other thing you may be expecting from this chapter is a load of information on what plugins you should use, what orders are most likely to work, how much reverb and delay to use, and so on. Whilst there will be a bit of this, there will be a greater emphasis on that part of producing an excellent vocal that most amateur producers neglect: Capturing the perfect recording.

The expression 'you can't polish a turd' has never been more aptly applied than in the context of vocals. It doesn't matter how good the performance is, how knowledgeable you are with subtractive EQ or serial compression, if the recording sucks. You cannot salvage a poorly recorded vocal.

So, the first thing I'm going to do is to hold your hand through every step you should take to ensure that you capture the best possible recording you can within your personal limitations.

The first tip is to keep your vocal recording dry. By this, I mean that, first, you should ensure that the space you're recording in is as free of natural reverb as possible. We'll look at how to do this shortly. Secondly, don't record any reverb or delay onto the vocal take. While the singer may want to hear some effects on their voice while tracking, ensure these are just for monitoring purposes. This is important for comping purposes first and foremost. Comping is simply the name for chopping together – compiling – different bits of different takes to make the one perfect take. If you have time-based effects burnt into your takes, you'll have conflicting tails, decays, and reflections all over the place. Equally, don't burn in any aggressive EQ or compression. Whilst you may end up cutting or boosting a lot of frequencies, it's better to have them all there in your project and have the option of what to manipulate in the mix.

Day 2 – Create space pockets

Let's consider the location in which you record. This location is the most influential factor in determining how your vocal will sound. It doesn't matter how good your mike is. If you put it in an awful-sounding space, it will simply capture more of the nasty characteristics of that space. So, wherever you record, whether it's a recording studio, a spare room, or even your closet, it's crucial to create a contained space in which to record your vocals.

Most amateur producers, and even reasonably successful ones these days, won't be recording in a space optimised for this purpose. Therefore, there are likely to be many reflections from walls and other surfaces that will colour your sound. These can be time-consuming, if not impossible, to remove afterwards with EQ and audio-repair tools.

An excellent option to combat this is to use a portable isolation booth. These provide you with a pocket of isolated dead space, allowing you to record a much cleaner performance. Isolation booths also bring the added bonus of forcing the singer to focus on their performance, as other distractions are blocked out. Use it to record anything in, not just vocals.

If a vocal booth is beyond your budget, there are other options. You could use curtains, blankets, duvets, pillows, or anything else you can get your hands on that will deaden your space. I've worked with a singer in Australia who recorded in his closet. He emptied it and covered the walls and ceiling with acoustic curtains. And he sounded great!

There will be a solution for almost any budget here. You just need the willingness to experiment and to try different things to get the most isolation you can out of the space you have available.

> TASK – Prepare your space to record vocals. It doesn't need to be expensive, although it can be if you want it to be! Do something that suits your budget.

Day 3 – Open-back vs closed-back

Getting the right sort of headphones for tracking vocals is fundamental to achieving a clean recording. There are two types of headphones on the market: Open-back and closed-back.

The rule for remembering which type you need for recording is simple: Closed-back closes everything else out, and open-back lets in other sounds. Therefore, for recording, you need closed-back headphones. You don't want the sound you're sending to your singer to bleed from the headphones and feed back down the vocalist's microphone.

There is a wide range of closed-back headphones on the market that cater to any budget. I'm not going to stick my neck out and say you should buy this or that model, as products on the market change. What I will say is do your research. Read reviews, read blogs, and watch comparisons from knowledgeable people on YouTube. And then make a decision that works for your budget.

I can hear you asking, 'But if I need closed-back for tracking, do I need open-back, too?' The answer is, no, you don't. If you can only afford one set of headphones, ensure they're closed-back. You can use these for tracking and mixing. Theoretically, open-back is much better and more comfortable for mixing as they provide a more natural, less isolated listening experience. However, if you're mixing in a space that isn't made for that purpose and therefore has ambient noise, then they're not that helpful. In a noisy working environment, you'll want to be as isolated as you can. Open-back is only beneficial if you're working in a space that is treated well enough to notice the benefit.

For the average amateur producer, one good pair of closed-back headphones is the way to go.[1]

> TASK – If you don't own a pair of closed-back headphones, you need to get some!

Day 4 – Use a pop shield

Always use a pop shield (also called a pop filter or pop screen). Don't ever skip this step! Even if the mike you're using has one built in, still use one. They're not too expensive to purchase, but you can make a makeshift one by stretching a pair of old tights over a coat hanger if money is tight.

The function of a pop shield is to stop plosives – that's t, k, and p (unvoiced) and d, g, and b (voiced) sounds – from getting to the microphone. The voice creates plosives when it stops the airflow using the lips, teeth, or palate, followed by a sudden release of air. The term plosive is onomatopoeic and is an abbreviation of explosive. A pop shield will prevent this rush of air from reaching the microphone, leaving you with a clean and controlled recording.

There are a couple of things to bear in mind. The first is size. There are different diameters of filters available on the market. You'll want one that's large enough to cover your mike's diaphragm. If your vocalist moves around a lot when they sing, then a larger diameter is better too. Secondly, you can purchase both nylon and metal filters. There are pros and cons to both. Nylon ones are cheap and great for removing plosives, but they can sometimes filter high frequencies and are easily damaged. Metal ones have wider holes, so they don't filter the sound, and they are reasonably durable, but it's easy to bend the metal sheet if you're not careful, and they can develop a whistling sound over time.

Once you've got your pop shield, you need to know how to set it up for optimum performance. The general rule is to position the shield three inches away from the microphone and position the singer a further three inches behind this.

> TASK – If you don't own a pop shield, buy or make yourself one.

Day 5 – Consider mike emulation

Assuming you've done everything you can to neutralise the reflections in your recording space, it's now time to turn your mind to capturing your vocalist's dulcet tones as best you possibly can. Of course, you'll want to use the best microphone you can get your hands on. But most amateur producers can't afford a top-end mike. It's easy to spend thousands on something high-end, which is well beyond the means of most hobbyists.

A good starting place is to get your hands on a solid all-rounder that won't cost you the Earth. We're talking condenser mikes here, not dynamic mikes. In general, they are much better suited to capturing the subtle nuances of a vocal performance. There are excellent, budget-friendly options from Røde, sE Electronics, Audio-Technica, and others.

After you've tracked your vocals into your DAW, you can then employ some mike emulation. This emulation will allow you to implement the characteristics of many expensive mike models for a snip of the cost. There are options for this from Waves, Antares, Universal Audio, and more. There are even bundles from Slate Digital, Antelope Audio, and others where you can purchase their purpose-built condenser mikes with all the mike emulations rolled together into one price.

With all these options, it's unlikely that you'll get the same result as using a five-grand Neumann, but with some experimenting, you'll get a very satisfactory result. The caveat here is that mike emulations will always work best if you spend a bit of time reading up on the characteristics of the mike emulations you're using. Sure, you can close your eyes and take a lucky dip, but it'll serve you better to make an informed choice.

> TASK – Do some research into mike emulations. Look at standalone plugins, as well as bundles. Consider if it's an option worth your investment or if you are happy using a single microphone.

Day 6 – DSP-powered monitor effects

You can assume that your singer will want to hear some effects and processing on their voice as they record. I don't think I've ever met a singer that doesn't! However, when you place processing and effects into your project as inserts, it takes some time for the sound to get from the vocalist into the microphone, into the DAW, through the processing chain, and back out into the headphones. This delay is called latency, and it's inevitable. You can reduce it by increasing your computer's processing power, reducing your buffer size, and carefully selecting low-CPU-using plugins. But it's still going to happen. And you can be confident that your vocalist will complain that they can hear a delay if you set it up this way.

So how do you solve this? The answer is an external DSP. DSP stands for Digital Signal Processing (confusingly, it also stands for Digital Streaming Platform!). What a DSP does is to process your audio externally from your computer. This processing will either happen within an audio interface with DSP built in or within a designated DSP unit or card installed into your machine. This external processing will allow you to throw whatever processes and effects you desire onto your vocal chain without worrying about running into the dreaded latency. If you record a lot of vocals, this should be a consideration when selecting which interface to purchase.

The most significant sticking point with DSPs is that they will typically only work with the plugins made by that manufacturer. So, the DSP built into Universal Audio hardware will only work with UAD plugins, the Waves external DSPs will only work with Waves plugins, etc. They're also expensive! So, you should ensure that you're confident they're necessary before splashing the cash.

> TASK – Research some DSPs. The market is changing all the time. I would recommend checking out UAD and Antelope Audio, to begin with.

Day 7 – The proximity effect

Lots of condenser microphones will have a low-cut filter built in. You should check your manual to determine what frequency it rolls off for your mike and how steep the slope is. It's well worth considering using this low-cut switch if you're working with a male vocalist and/or your singer is performing quite close to the mike. This low-cut switch will help reduce the proximity effect. The proximity effect is a phenomenon that causes an increase in low frequencies as you move the mike closer to the singer. The closer you get, the more of a bass boost you'll get.

On the face of it, this may sound like a bad thing, but it has pros and cons. The proximity effect is what allows radio DJs to have that stereotypically thick, rich tone of voice. It's also allowed many singers through the years to enrich their voices beyond the realms of realism. Conversely, it can easily make things less intelligible and muddy and make vocals get in the way of bass instruments. So yes, you can use it to fatten things up, but you need to be careful.

Using your low-cut switch will help you avoid some EQ'ing later. With that said, there's nothing stopping you from applying this high-pass filter in your DAW post-recording. The benefit of doing it this way round is that you can adjust the filter frequency and slope. There isn't a right or wrong way to do something, just personal preference, as with most things.²

> TASK – Explore the proximity effect with your microphone. You should be able to identify what it sounds like so you can avoid it or implement it as you wish.

Day 8 – Get your levels right

At this point, you're almost ready to hit the record button. But there are two more essential hoops to jump through before you do.

The first is to ensure that you've got your input levels right on your vocalist's mike. Do several test recordings, ensuring that your singer goes as loud as they're going to in the 'real' takes so you can ensure you're not going to overload on input. Be cautious here rather than reckless. It's better to boost a slightly quiet signal than to try to salvage a clipped one.

If you're uncertain, or whilst you're learning how to get this just right, there are plugins out there that have a built-in clip-safe mode.

While we're on this subject, is there a golden number you should aim for on input? Not exactly, no. A quick Google search will provide you with a range of answers, but they're all loosely in the same ballpark. So, my ballpark is this: Aim for an average level of around −14 to −18dB peak, not exceeding about −10dB, but absolutely no more than −6dB, and going no quieter than around −20 to −24dB. Vague enough for you? Great! Let's move on.³

> TASK – Practise trimming your microphone. You can do it with your own voice or on someone else's. Look to get the level peaking in that sweet spot around −14dB.

Day 9 – Get your mix right

The second hoop you need to jump through is to get your vocalist's mix just right. The more comfortable your vocalist is with their recording environment, mix included, the better the performance they'll give you will be. What interface you have matters a great deal at this point. You'll want to set up a personalised mix for your vocalist that is different from the current static mix in your project.

Many interfaces these days come with software that allows you to route things as you want them, but you'll want to research this before purchasing to be safe.

For the sake of argument, let's assume that your main mix is coming out of outputs 1 and 2 and that you've set up the vocalist's headphone mix to come out of outputs 3 and 4. Within your project, you'll then want to set up a new buss and assign the output of that buss to outputs 3 and 4. Label it as 'Vox Mix' or something logical. You can then send whatever amount of any element in your track your vocalist desires to this new buss without altering your actual mix at all. Whilst you're at it, set any of these buss sends to pre-fader. This will mean that, should you want to tweak your mix whilst your vocalist is tracking, it won't affect what the singer is hearing. Adjusting your singer's mix whilst they're performing will be distracting, so having their unique mix on pre-fader sends is essential. The last thing you want is a distracted singer. You want them to be in the zone, 100% focused on giving you the very best they're capable of.

> TASK – Practise setting up Vox Mixes on a new buss going to an independent output. Ensure your sends are set to pre-fader.

Day 10 – Record *everything*!

Working with a singer is different than working with an instrumentalist. Instrumentalists can generally go for hours and hours without getting too tired. On the other hand, singers can't. Their voices get tired. Therefore, you have a finite window in which to get what you need out of them.

For this reason, it's common practice to record everything they do. When you're getting your vocalist's headphone mix right, when they're warming up or doing a practice run, record it all. Whether you choose to be transparent about this with your vocalist is up to you.

You will often get a completely different performance when the light is off instead of when it's on, figuratively speaking. Vocalists are often more relaxed when they think the pressure is off. They can often become tense or overstretch in pursuit of perfection when doing the 'real thing'.

This concept isn't unique to amateur vocalists. The same has been used countless times with professionals, too. The beauty of working in the digital domain is that your computer will come with a large amount of storage space, and you can buy additional space relatively inexpensively. You can also create as many tracks as you like within your session; you're not limited by the number of channels on your desk.

One of the most credible skills to master as a producer is to be able to identify good bits of takes in real time. It could be a line from warm-up A, two lines from take two, a phrase from warm-up B, and that long note from take five. It's good practice to get your singer to provide you with a printed copy of the lyrics for your session. You can use these to mark up good bits of various takes as you go. Doing this will save you from having to listen back through everything afterwards. You can just hone in on the bits you already know you liked the sound of. Your singer is also likely to be impressed by this. Time is money. The less time you spend trawling back through takes, the more time you have to produce the track.

> TASK – If you're not doing so already, I would recommend practising with a singer at this stage. Record a few takes of a song or section and practise identifying the best bits of each performance in real time.

Day 11 – Harmonies and layers

A lot of plugins on the market are designed to thicken a vocal. Their marketing says things like, 'Create immediate double tracks'. Whilst you can get some reasonable results out of them, I'm curious as to why you'd ever want to use them. To me, a vocal that has been artificially created can be spotted a mile away. It's hard to explain, but even with the most precise formant shifting, tuning, chorusing, and every other trick you can throw at it, it still doesn't sound natural. Of course, sometimes, this may be precisely what you want. In these instances, great! Knock yourself out!

But when you're set up with a vocalist right there with you in the 'studio', why wouldn't you use them to lay down any harmonies and layers that you want? It just makes sense!

I will always get my singers to put down a double track of the lead vocal and usually a triple track. Having these options for your mix is great. I will often keep the lead vocal on its own for the verse, perhaps layer the double track for the pre-chorus, and then add the triple as well for the chorus. With just the double track in, I'll generally layer this directly underneath the lead, without panning, and then when I bring the triple in, I'll pan both off centre a little to widen the lead vocal.

Something I'll also generally track is a lower and higher octave of the lead. This is assuming that your singer can reach the notes. They don't need to be performed at the same dynamic as the lead, so long as there's still conviction. But they're great to have to layer into your mix subtly. Almost not being able to hear them is usually perfect.

After these layers, gather whatever harmonies you and the singer want to put down. Generally, the more the merrier. It's much better to have lots of options and creatively select what you want to use in your mix. I recommend getting at least two versions of each line. You can then hard pan these in opposition, creating wide backing vocals.

Finally, your vocalist may have specific ad-libs they wish to interject. Usually, they'll have loads of ideas and will want to put them all over the place. That's fine, let them! Again, you can then select the few that will make the cut in the mix. But the more options you have, the better.

> TASK – Get used to capturing all the vocal layers: Double track, triple track, low and high octave, harmonies, and ad-libs. Focus on keeping your session well organised, with each part on its own track and everything labelled appropriately.

Day 12 – Gate post-recording

Noise gates are things that you use to cut out unwanted sounds from a track. They have similar controls to a compressor. They have a threshold, which is the point at which the gate will open and close, and attack and release controls, which denote how fast the gate opens and closes once the threshold is exceeded. The difference is that the signal above the threshold can be heard, whereas the signal below is cut out. On the face of it, this is a great thing, especially if you're working in a less-than-perfect environment where there may be background noises.

The main problem here is that a live vocal is an organic thing. The dynamic of the performance won't be the same throughout; there will be dynamic variety. But a gate's threshold is a fixed point. Sure, you can automate it, but I can't imagine a time when this would ever be worth the effort. And with your threshold fixed, some parts of your vocal are likely to get cut off as they fall below the required level to keep the gate open. You don't want this!

If you want or need to use a gate for whatever reason, I recommend doing it post-tracking. Record everything that you need without any gate and add the gate afterwards. I hardly ever use a gate to cut unwanted sounds out. I don't like the risk involved in potentially cutting out something I want to keep. I will always manually edit out the bits of regions I don't want to keep.

Working this way keeps me 100% in control of what I hear and what I don't. Most DAWs will have a function called 'remove strip silence' or something similar. This function is a sort of middle ground between the noise gate and the manual method, where you can automatically cut out silence from your region based upon a threshold and length of silence that you specify.

Another option if you're set on using a gate is to use the reduction function. This will usually be labelled as a percentage. It means that once the signal dips below your threshold, it will be reduced in level by the percentage you specify. In this way, you can avoid completely cutting stuff out.

> TASK – Explore noise gates on your recorded vocal. Most likely, your DAW will have one as stock. See if you can dial in the sweet spot so the gate opens and closes naturally between sung phrases.

Day 13 – Loop and comp

It's almost impossible to record the perfect vocal performance in one go. Yes, it is possible in theory, but in the modern music climate where your audience expects nothing short of flawless perfection, it's standard practice to splice together bits of multiple takes that work together. This process is called comping.

Comping multiple takes together is not a reflection of the singer's ability to deliver. It doesn't mean they're incapable of providing you with an outstanding performance. But why wouldn't you maximise the value you get out of countless numbers of takes without the worry of the expense and the additional editing time of cutting them together? This was what happened in the analogue world. Comping in the digital age is so easy; you'd be foolish not to. That is unless what you're going for is a raw, edgy performance that is clearly delivered in one take.

A standard approach to take is to loop a specific section at a time and go round it until you feel you've got what you need. And then move on to the next part. This approach has a couple of benefits. First, it allows you, the producer, to pay more attention to each take compared to the ones before and after it and make notes on potential winners. Secondly, often different sections of songs have contrasting dynamics. Verses are usually quieter than choruses. From a mixing perspective, it can make sense to split these sections onto other channels. So, you have a verse vocal channel and chorus vocal channel, and so on. You'll probably want different compression levels on each, perhaps different EQ settings to compensate for the contrasting dynamics, different effects levels, and so on. By utilising different tracks, you'll save yourself time in the automation department down the line.

> TASK – Find out how to, and practise, comping in your DAW. All DAWs will allow you to comp in one way or another. You'll want to find some content online that is specific to your DAW to learn this.

Day 14 – Tuning and other modulation effects

Once you've comped all your vocals, and before moving on to EQ, compression, and everything else, you should do all your editing. There are three stages to this.

The first is to edit your vocals' timing. Some DAWs have functions built in that you can use; others don't. Your objective here is to tighten up your vocals' rhythm. How much or how little you adjust here depends on the genre and personal preference. You may not want to edit the timing at all, or you may want to put everything totally on the grid. Neither approach is right or wrong; it's subjective. This stage is also your chance to tighten any doubles and harmonies. You may want your backing vocals to be super tight to create a slick, highly polished feel. Or you might not. Again, some DAWs have built-in functions that will help you use one track as a guide to adjust the rhythm of another; others don't. There is an excellent plugin by SynchroArts called VocAlign that makes light work of this otherwise time-consuming process.

The second step is to tune your vocals. When comping vocals, I never worry too much about tuning. I know I can tighten up any tuning issues with modern tuning plugins so long as they're in the ballpark. I'm always more bothered about timbre, energy, and dynamics. Again, the tuning functionality built into DAWs varies significantly, but there are many great third-party options for this. The industry standards are from Celemony and Antares. Once again, how aggressively you tune your vocals is a personal choice.

After these two steps, the final thing to consider is modulation effects, primarily vocoders and autotuners. I have deliberately put this step after all the other steps in the vocal process. In amateur productions, autotuners or vocoders will often be slapped onto a channel right at the beginning before a note is sung. This will remain on throughout, meaning the dry vocal will never be heard. So, neither the producer nor the singer will ever know if the raw vocal is any good. Autotuners and vocoders work so much better when a quality vocal performance is fed into them. Putting an autotuner on a well-edited, well-tuned performance means that the plugin doesn't have to work so hard to snap notes to the grid, and therefore you'll get fewer glitching artefacts and oddities.

TASK – Learn how to adjust timing and tuning in your DAW. Some DAWs have excellent timing and tuning functionality built in. Others don't. In which case, you may wish to consider investing in a plugin to assist with this process. Do your research and make an informed decision.

Day 15 – Subtractive EQ

Once you have your vocal perfectly edited, it's time to begin working it into your mix. The usual starting point is with subtractive EQ. You should be listening to the tonal qualities of your vocal here, specifically, aiming to tame or remove any problem frequencies.

A typical starting point is to remove excessive low-end build-up. The human voice doesn't have much valuable frequency content below 80–100Hz, so you can safely roll this off with a high-pass filter. This is more likely to be necessary for male voices than female, where the proximity effect is expected to be more of a factor. The exact frequency that you set your filter at will depend on the singer's voice and the pitch at which they are singing. It's common to select one cutoff frequency and apply it across all your vocal channels, but this is lazy. There will be more low-end content in a lower passage, so a lower cutoff frequency may be more appropriate.

You should also look out for any harsh frequencies. Some people will advise you to grab a narrow band, boost it, sweep it until you hear a frequency that sticks out more than the rest, and then cut this out. I find this approach heavy-handed and it can leave you with a soulless vocal if you're not careful. If you've taken the care to get your recording right in the first place, you shouldn't need to cut harsh frequencies too aggressively. Instead, you can still grab a narrow band, sweep it until you hit a pokey frequency, but then use a dynamic band to attenuate it gently rather than cutting it completely. This will mean the frequency is only being acted upon when it exceeds that band's threshold, which will be much more pleasant and far less intrusive.

As a reminder, for subtractive EQ, it's good practice to use a transparent equaliser. Your DAW's stock EQ will likely fit the bill, although it may not have dynamic bands. Options that I turn to are FabFilter's Pro-Q3, Slate Digital's Infinity EQ, and Waves' F6.

As a final reminder, use narrow Qs when you're cutting frequencies. You only want to cut the offending frequency, not those around it.

TASK – Spend some time practising subtractive EQ on vocals. This is one of the most challenging skills to master, so don't give yourself a hard time if you can't nail it immediately.

Day 16 – Consider the genre

Once you've removed all the unnecessaries from your vocal, it's natural to want to dive straight into tone-shaping. Before you do this, I implore you to think first about the style of music you're making.

The genre you're producing will often dictate specific stylistic characteristics that will make your vocal work more convincingly within that setting. Here's a quick rundown of some of the standout points:

- For pop, RnB, and most electronic music, you will be heavily processing the vocal. You'll want lots of top-end shimmer, clearly audible processing, and a consistent dynamic throughout.
- Hip-hop is like pop, but with less top-end shine and fewer effects. You'll want more aggression and presence in the upper mids and a heavy low end to provide more power.
- For rock music, you've got less top end but more high mids and body. Your vocal will generally sit a little deeper in the mix too.
- In jazz, subtlety is the name of the game. You won't want any noticeable processing and will want to keep all your dynamics intact.
- Metal and hardcore use a lot of heavy compression, which helps achieve the distinctive, aggressive tone. It'll have less low end, more body and high mids.

Only by thinking about the sort of music you're making at this stage will you be able to make solid decisions about how you will process your vocals. This also requires you to know a bit about the stylistic features a listener will expect from the music you're producing. The more information you're armed with, the more likely you are to get the production just right.

> TASK – Compare the vocal production in some contrasting genres. Can you hear any of the characteristics I outlined?

Day 17 – Tone-shaping EQ

Now that you've got your head in the right state of mind for the style you're working with, you're ready to start shaping the tone of your vocal. I've deliberately labelled this as tone-shaping rather than additive EQ because shaping the tone of your vocal doesn't just mean that you're going to be enhancing the things you want to hear more. As you may have assumed from the implications in yesterday's tip, this stage may also call for some subtractive EQ. The main difference between the subtractive EQ here and before is that you're not looking to remove unpleasant things this time. Think of yourself now as a sculptor working with a beautiful piece of wood. The wood grain is gorgeous throughout, but that doesn't mean you need to use it all. You're still going to carve away parts of it to enhance the bits to which you want to draw attention.

Therefore, when making tone-shaping EQ moves, you'll want to apply broader brush strokes. This means using a wider Q. Keeping your bandwidth wide will maintain the natural qualities of your vocal. In Unit 3 we talked about using analogue EQs for tone-shaping. On some analogue EQ models, you won't even find a Q control to utilise. A Neve 1073 or an API 560 are great examples of this.

On the other hand, SSL models do have bandwidth controls. What you use is down to personal preference and the sort of tone that you're looking to get out of your analogue emulation. Some professional engineers will always use the same analogue EQs, regardless of the genre within which they're working. This may be because that's the console that they have in their studio, or it may be that they just feel comfortable working with that model. There aren't many rights or wrongs in the music-production world, as you know by now. You can often plot many different routes to reach the same location.

> TASK – Practise tone-shaping the same vocal, imagining it across multiple different genres. What would you do differently from style to style? Try it with different analogue EQ models if you have them. Does the model make a difference to the overall vibe of the vocal?

Day 18 – Compression

As previously mentioned, the style of compression you use and how much of it you apply will be completely different, dependent upon the genre. Jazz and hardcore are at opposite ends of the spectrum. What I'll do here is take you through some standard techniques. How much of this guidance you choose to apply is up to you.

You already know that different compressors have contrasting tonal characters. What you choose to use will depend on how you want it to affect your vocal. Using a 1176 or SSL-style compressor first in your chain is common. A good starting point is to dial in 3–6dB of gain reduction. You don't want the compressor to be working all the time. It should just be engaging on the loudest peaks.

The attack and release times are critical. To begin with, dial in the slowest attack and fastest release times. Increase the attack gradually until you start to shave the transient off the vocal and then back off. Then decrease the release until your compressor is breathing in time with your track.

You don't want to be compressing all your vocal signal at this stage. It's OK for your needle to return to zero before engaging on the next transient peak. You don't want to squeeze the breath out of it, but you *do* want to add energy and power.

Whilst you're learning to hear this, I would advise using a compressor with a visible needle that will bounce. A 1176 gives you this. Just be careful, because if your 1176 plugin is built to model the original hardware truly, the attack and release dials will work backwards. So, when you turn the dials up to 7, this is the fastest time, and as you back off towards 1, they get slower. This is just a quirk of the original outboard gear that most manufacturers choose to emulate.

> TASK – Practise this first stage of vocal compression on various vocal performances. This technique is most applicable to faster-paced, more transient-heavy performances, so try to select appropriate material to use.

Day 19 – More compression

It's usual to compress vocals in series. As a reminder, this means using multiple compressors one after the other. It's normal to start with a fast peak-limiter-style compressor like the models I mentioned yesterday to rein in the transient peaks of the vocal. You will often follow this with a slower compressor that will act upon the main body of the vocal, levelling out the dynamics.

I recommend you use either an Optical- or Variable Mu-style compressor for this job. Some sort of LA-2A or Fairchild 660 emulations are my favourite. The amount of compression you're looking for should be the same as with your first compressor: Around 3–6dB. Compressing gently in multiple stages like this will mean that you don't notice that your signal is being obviously compressed. However, it will be noticeably flatter in dynamic, making it much easier to bed into the front of your mix.

While learning this second serial compression stage, I recommend using an LA-2A emulation. My reason is that they're straightforward to understand as they only have two dials. Like the 1176, they have their own quirk, this time in how the signal flows. You'll find the peak reduction knob on the right-hand side, which you should use first to set your compression amount. You'll then use the gain knob on the left to match your input and output signal. As you've already learned, the LA-2A is a musical-sounding compressor that maintains all the natural qualities of your track whilst providing the consistency in dynamic that you're after.

> TASK – Implement serial compression on a range of vocals, using an optical or Variable Mu compressor after a peak-limiter compressor. Experiment with different compression amounts to see how much you can get away with whilst keeping it natural.

Day 20 – . . . and even more compression!

In modern music-productions, where everything feels polished to within an inch of its life, it is standard practice to use parallel compression.

The idea here is to blast your vocal, destroying all dynamics to make an entirely flat, consistent signal. As it's in parallel (meaning it runs alongside your primary signal), you can blend a tasteful amount of it underneath your main vocal track to provide additional security and consistency.

You should be using a high ratio, fast attack and release times, and a low threshold. Don't worry if you're achieving 20dB of gain reduction here; that's the point.

Regarding models you may want to use here, another 1176 will work fine. But there are some more aggressive options that you can employ. The hardware Empirical Labs Distressor is great for this, and UAD and Slate Digital both make excellent plugin versions. Alternatively, from Soundtoys, you've got Decapitator and Devil-Loc, both of which can be impressively aggressive and in your face.

> TASK – Experiment with some parallel compression. If you have different compressor options, explore them all. Do you prefer the character of one to another? Does it depend on the genre, or the vocalist?

Day 21 – De-ess with care

Sometimes vocals will have some harshness to them caused by excessive sibilance. When this happens, you'll want to use a de-esser to tame it. Note that this won't always be the case. Several things will affect how strong the sibilance is in a performance: The way the vocalist enunciates, the dynamic at which they're performing, the rhythm of the vocal, and even things like the choice of microphone, pop shield, and distance from the mike can all play a part. Use your ears to determine if de-essing is necessary or not.

Where in your signal chain you should de-ess is subjective. Some producers like to do it at the start of their chain or just after subtractive EQ, some after compression. There isn't a right or wrong. To my mind, doing it after compression means the sibilance will have become more evident due to the levelling out of the dynamics in the performance via the compression, which should make it easier to identify and target.

A de-esser is simply a compressor that targets a specific frequency range. A good one will have a monitor feature built in. This monitor feature will allow you to listen to the affected frequencies only, meaning you can dial in just the right frequency range and reduction amount. Every manufacturer makes one, but the one that comes as stock in your DAW is probably OK too.

The vital thing with de-essing is not to be heavy-handed. Going too hard with it will leave you with a vocalist that sounds as if they have a lisp. You don't want to eliminate the 's' and 't' sounds, as this would leave you with an inarticulate, unnatural-sounding performance. Be gentle and ensure that the vocalist sounds human.

> TASK – Practise de-essing without going too far. Put the de-esser in different positions in your signal chain. What difference does this make?

Day 22 – Saturation and distortion

The term saturation has become very trendy, but it's massively misunderstood. Saturation is distortion applied subtly. A little-known fact is that overdrive is the same. It simply sits in between saturation and distortion in terms of intensity.

Saturation traditionally came from analogue recording setups and came about through preamp gain, console circuitry, and tape machines. These devices would enrich the audio signal in one way or another by adding additional harmonics to the frequency content, making the signal sound thicker. Every device would colour the signal differently depending upon the components used. Therefore, vintage audio equipment is highly sought after and very expensive. Often, hardware devices were produced in limited runs and then discontinued. They were subsequently found to have a particularly pleasing sound, and thus their value increased exponentially. The original UREI 1176 A compressor of 1966 is a prime example. Only 25 were initially manufactured, and there have now been nine revisions and countless other companies making their version of the 1176.

Fortunately, all these things are available via emulations in the digital domain, and each will have its own characteristics. The differences are subtle and take a tuned ear to identify. The best way to learn is to drive your sounds hard through a preamp, console, tape machine emulation, etc., respectively, to see what they're all doing.

Saturating a vocal can make it feel brighter and more exciting. It can help it hold its own more within your mix and make it feel thicker. How you apply the saturation and with what is up to you. You can place it in series or parallel, as subtly or aggressively as you wish.

> TASK – Experiment with different saturation options. Explore everything you have (preamp, console, tape, distortion, etc.) and see if you can hear their differences. Try them in both series and parallel.

Day 23 – Don't overdo reverb and delay

After all these steps, you'll have a great-sounding vocal to use in your mix. The vocal's dynamics and tone will be just where you want them. You'll now probably want to reach for some time-based effects such as reverb and delay. These effects will provide your vocal with a sense of space. When thinking about your space, consider it along two axes. First, the x-axis spans from left to right across the width of your stereo field. Secondly, the z-axis is your depth axis. Picture this like a long straight road that disappears into the distance in front of you.

Stereo delays and reverbs help add a sense of width, whilst mono effects will work better for creating depth.

As we've explored previously, the longer the time of the effect, the larger the perceived space will sound, and vice versa. It's common to set their lengths to match the tempo of your track. If you want a clear vocal, time your effects so that they decay before the beginning of the following phrase. This will avoid making your mix muddy. Conversely, if you want something washy, you'll want to ensure the decay overlaps with the vocal phrases.

One of the most common problems I hear in mixes is that the time-based effects have been applied too liberally and are getting in the way of other mix elements. These time-based effects should generally be implied, not overstated. A helpful method I have used for a long time is to increase the effect amount gradually until I can just hear it within the context of the mix and then back it off a few dB. This will ensure that it's there but not overly prominent.

> TASK – Implement contrasting time-based effects on some vocals. Explore stereo and mono effects, as well as different lengths. Importantly, try to get your effects' level just right so it doesn't overpower your mix.

Day 24 – Pan backing vocals

It's common knowledge that your lead vocal should stay down the middle of your mix 99.9% of the time. But where to pan backing vocals is much less clear. Think about them creatively rather than trying to follow a predefined set of rules for your BVs.

Here are some informative questions that you should ask yourself:

1. What is the BV doing? Is it doubling the melody? Is it harmonising? Is it singing actual words or performing oohs and aahs?
2. Is it working congruently with the lead vocal or interjecting antiphonally?
3. Is it a countermelody that should be given equal importance, or is it playing a supplementary role?
4. Is it an ad-lib, or is it a carefully written part?

By identifying the sort of BV you have in front of you, you'll be able to make a more informed decision about where it should be panned within your mix.

As a rule, the more fundamental the vocal is, the closer to the centre of your mix you should have it. For example, having a countermelody panned hard out to the left would sound odd.

However, double-tracked BVs are different beasts. You can pan these hard left and right in opposition to provide a wide feel to your mix. Adjusting the width of these elements throughout your mix will create contrast.

In general, automating the position of BVs is a creative way to make your mix more interesting. Having one thing static in your track from start to finish can get stagnant after a while. Consider moving things around subtly to keep things fresh.

> TASK – Revisit backing vocals in a few mixes. Ask yourself the previous four questions and see if their positioning is appropriate. Make relevant adjustments.

Day 25 – Converting to MIDI

A nifty feature to play around with is converting audio to MIDI. Most DAWs have some version of this functionality built in somewhere. It allows you to extract the pitch and rhythmic information from an audio track and create MIDI information from it. This process comes with a warning: The more complex the audio content is that your DAW needs to analyse, the less successful the MIDI conversion is likely to be. The resulting MIDI is likely to be far from perfect if the vocal performance has lots of grace notes, wide vibrato, glissandos, and portamentos, etc. The simpler the audio is, the more accurate the MIDI conversion will be.

However, it's possible to go into your piano roll editor and tidy up the resultant MIDI content. This shouldn't cause you too much of a headache.

The critical question is, why would you want to do this? And the answer is simple: It will allow you to layer other instrumental sounds underneath your vocal to support it. Maybe you want a synth doubling your lead vocal an octave higher; perhaps you want to bolster your BVs with some synth choir sounds to thicken them up. There are lots of creative ways that you can use this tool.

The other thing that this is good for is creating sheet music. By converting your audio to MIDI, you can then use it as your starting point for creating a score. Professional programs such as Sibelius and Nuendo can successfully import MIDI. Doing your tidying up in your DAW first and then transferring the MIDI over will save you a lot of time. This is a pretty niche scenario that won't apply to most, but it's worth considering for some.

> TASK – Find out if your DAW has audio-to-MIDI functionality. If it does, learn it. What creative uses can you find for it?

Day 26 – Vocal automation

As we know already, automation is the best way of keeping your mix moving, making it feel as if it's a living, breathing thing and not something static. Useful vocal automation devices are worth revising here as your vocal is likely to be so prominent in your mix that automating elements on it will make a significant difference to your mix.

The obvious place to start is by automating the volume of your vocals. Not everything needs to be solved with compression. Too much compression can be detrimental overall as it can remove the expressiveness of a performance. Riding the level manually with automation is great for keeping your vocal level consistent or just to help some sections pop more than others.

Sticking with levels for a moment, automating the amount of parallel compression you have present in your mix is excellent for providing fuller sections with the consistency of level required to balance with everything else.

Changes in delay, reverb, and saturation are all valuable things to automate if you're looking to develop your mix subtly or if you want to make a word or phrase stand out more from others. As with most things, subtlety is paramount here. Automating heavy-handedly is likely to do more harm than good. You should be aiming to take your listener on a journey without blatantly signposting every change in direction.

> TASK – Revisit a few vocals. If you've automated them already, consider whether you're over/underdoing it. You want your vocal automation to enhance the track without drawing attention to itself.

Day 27 – Frequency allocation

The final piece in the vocal jigsaw puzzle is frequency allocation. Everything you've done to this point is focused on the vocal itself rather than its relationship with other mix elements. It is almost certain that other parts in your track will be masking your vocal as they share the same frequencies.

The answer here is to allocate frequencies to your vocal. This is another massively misunderstood concept that really shouldn't be scary. The process is simple: You decide which frequencies are most important to your vocal, identify the other conflicting parts in this range, and gently carve these out, leaving space for your vocal to shine through. Let's look at a couple of examples of this.

The first vital frequency to make space for is the vocal's fundamental. The fundamental frequency is the lowest bump in the frequency spectrum. You'll want to use a frequency analyser to find this. Voxengo's SPAN is free, or you'll find it built into some EQs such as Fab Filter's Pro-Q 3. You'll typically find it between 80–180Hz in male voices and between 160–260Hz for female voices. Reducing the level of this fundamental frequency in competing instruments will allow the true tone and body of the vocal to shine through. This is a great place to look if you think your vocal sounds a bit thin and flimsy.

Identifying other frequencies to allocate can be a bit of a mystery, so here's how I like to look at it: When I do my tone-shaping EQ, I may well boost some frequencies in my vocal that I like. It's good practice to do this EQ'ing in the context of your mix and not in isolation. However, you may find that you must push it aggressively to get the specific frequency that you like to cut as you desire. Rather than boosting 4kHz by 8dB in your vocal, try boosting by 4dB only. Then find other competing instruments in this range and cut them by 4dB to carve room for your vocal. This method is likely to keep things sounding more natural across the board, which is probably what you want.

If you're feeling particularly lazy or just lack confidence with this, you can use a plugin to help with this process. Trackspacer by Wavesfactory is the standout in this area.

Let's caveat all this by saying that you only need to do frequency allocation if required. Don't go looking for problems that aren't there. With all processes, use your ears and intuition to make an informed decision about what is needed.

TASK – Practise frequency allocation with vocals. Find instruments that are getting in the way of your vocals and make space for them. Sidechained dynamic EQ is a great tool to use here.

Day 28 – Unit summary

Hopefully, now you feel much more informed on the process of capturing and processing vocals. As a quick reminder, here's what we've covered:

- Creating a space in which to record vocals
- Open-back vs closed-back headphones
- Pop shields
- Microphone selection
- DSP-powered processing
- The proximity effect
- Getting levels right first
- Recording everything
- Harmonies and layers
- Gating
- Looping and comping
- Time and pitch correction
- Considering the genre
- Subtractive and tone-shaping EQ
- Compression, compression, and even more compression!
- De-essing
- Saturation
- Time-based effects
- Panning BVs
- Automation
- Frequency allocation

I hope that now you've scrubbed the concept of a stock channel strip for vocals from your mind. As you now appreciate, every vocal will have its own set of characteristics and requirements that cannot be catered to by something premade. There are so many factors that affect the vocal's quality that you must consider that trying to apply a 'one-size-fits-all' approach is doing your vocal a disservice.

We're now very close to the home stretch of this course and the unit that many of you have been waiting for: Mastering. But before we get there, I want to cover what many consider the next most daunting topic: Synthesis.

Checklist

- Have you optimised your recording environment as best you can?
- Do you have the right headphones for recording?
- Have you got a pop shield?
- Is your microphone's positioning correct?
- Are your recording levels good?

- Have you got your singer's mix right?
- Don't forget to record *everything*!
- Have you captured additional layers, harmonies, and ad-libs?
- Have you comped your vocals?
- Have you edited your timing and tuning as you wish?
- Have you considered the stylistic characteristics of your vocal?
- Have you removed unnecessary frequency content and shaped the tone with reference to the genre?
- Have you applied a suitable number of layers of compression considering the genre?
- Have you de-essed, if necessary?
- Have you applied saturation, if necessary?
- Have you panned your vocals considering their independence from the lead vocal?
- Have you automated elements of your vocal to maintain interest?
- Have you implemented frequency allocation to help your vocal cut through the mix?

Having this checklist by your side whilst you're implementing the steps into your workflow will help you.

Further reading

1 Koester, T. (2022). *Open-back vs closed-back headphones: What's the difference?* [online] sweetwater.com. Available at www.sweetwater.com/insync/open-back-vs-closed-back-headphones-whats-the-difference/ [Accessed 9 Nov. 2022].
2 DPA Microphones. (2022). *Proximity effect in microphones explained: How it affects different sound sources.* [online] dpamicrophones.com. Available at www.dpamicrophones.com/mic-university/source-dependent-proximity-effect-in-microphones [Accessed 9 Nov. 2022].
3 Wregleworth, R. (2022). *What dB should vocals be recorded at and why?* [online] musicianshq.com. Available at https://musicianshq.com/what-db-should-vocals-be-recorded-at-and-why/#:~:text=You%20should%20record%20vocals%20at,it%20comes%20to%20recording%20vocals%3F [Accessed 9 Nov. 2022].

Unit 9

Synthesis

Day 1 – Synthesis terrified me

Let me be transparent with you from the outset. Synthesis terrified me for many years. It was always that one gaping hole in my knowledge that I was embarrassed to acknowledge. And I know I'm not alone in this situation.

So many producers have an excellent understanding of almost all elements of music-production, except for synthesis. For some reason, it scares people. Maybe it's the overwhelming feeling you can get when you first look at a synth. Whether it's a hardware or software model, being confronted by so many dials, knobs, and buttons is enough to put you off for life. Maybe it's the geeky taboo that is associated with synthesis. If you were in this camp like me, let me change your mind.

Don't misunderstand me. I'm not going to turn you into a synthesis guru during this next chapter. I may not even make you like synthesis. I'm not even sure that I do.

What is certain in my mind is that all producers worth their salt should have a solid, basic understanding of synthesis, how it works, and how to manipulate it. Synths are used in all genres of music these days – everything from death metal to film scores. So, if you think you're going to avoid them, you're gravely mistaken.

I am not aiming to teach you how to design the most mind-blowingly complex patches from scratch. That would take far more time than I have here. And I'm not the best person to teach it to you. However, you should be able to take a preset and understand what you need to reach for to dial it in perfectly so that it beds into your track just the way you want. As you know, I don't subscribe to the 'just fiddle until it sounds right' school of thought. That method is a giant waste of time. I want you to be able to listen to a synth sound and identify what you need to reach for to adjust it to your liking.

During this unit, I will do my best to remove the mystique surrounding the synthesis world. And who knows, by the end, perhaps you'll even be inspired to dive a little deeper into the subject.

DOI: 10.4324/9781003373049-9

Day 2 – What is a synthesiser?

Before we get into what a synthesiser is, it's crucial to understand how we as humans hear sound. When something vibrates, whether it's a string on a guitar, a head on a drum, or anything else, it makes the air around it vibrate too. These air vibrations travel into our ears, where they're converted back into sound by our brains. This is why you can't hear sound in a vacuum. If there's no air to transfer the vibrations, there's no way of it entering your ears.[1]

A synthesiser replicates this process, but rather than physically vibrating a string, a skin, or anything else, the vibration originates in an electrical signal. This electrical signal can be slowed down, sped up, and manipulated in ways that acoustic sounds cannot. The signal will still need to be amplified and sent out through a speaker so that it can vibrate the air that will ultimately enter your ears.

You may feel that synthesis is an entirely unmusical process to this extent. Creating something entirely digitally doesn't sit well with some. But consider this: Anything you record becomes altogether digital when it enters your DAW. My limited understanding of the inner workings of computers is that everything can ultimately be boiled down to a series of noughts and ones. This isn't so far removed from synthesis, then.

Consider this, too: Every natural sound created has a unique timbre. This is what allows you to know that you're hearing a violin and not a tuba, for example. Synthesis can be interpreted as observing the characteristics of a waveform created by, for example, a string section and trying to recreate it artificially. Not convinced? Look at it this way. Imagine you need a heart transplant (please forgive the extreme example). You have two options: Option one is that you wait for a perfectly matched heart to come along, which will come at vast expense, if it comes at all. Option two is to take a synthesised heart. The artificial heart will do the job well, to the untrained eye, almost convincing you that it is, in fact, real. And it comes at a snip of the cost of a real one. Now consider that the real heart is a live string section, and the synthesised one is your plugin instrument. Point made.

> TASK – Look at the synths that you have. Most DAWs will come with a range of options. Try to identify common features between them. Study their layouts. Identify what is different between them.

Day 3 – Oscillators

Let's begin to learn how to synthesise. A synth is nothing without an oscillator because the oscillator is what creates the electrical signal. Without it, there is no signal to be manipulated. So, this is your starting point.

Oscillators put out a repeating waveform. It's called an oscillator because the waveform literally oscillates above and below the zero-crossing line. These waveforms are regular and predictable.

There are three ways in which you can manipulate an oscillator. First, frequency. These frequencies are the same as those you're familiar with on your equaliser. The frequency (measured in Hertz) is the time it takes to complete one waveform cycle. The longer it takes, the more elongated the waveform is, and therefore the lower the pitch is. Conversely, the tighter the wave's cycle is, the higher the pitch is. How you adjust this will almost certainly be determined without thought as you play the oscillator with your MIDI keyboard. The note you play will instruct the oscillator of the frequency you want the oscillator to operate.

The second way of manipulating the oscillator is through amplitude. Amplitude is simply loudness. So, the amplitude is literally how loud the wave is when it leaves the oscillator. This amplitude is measured in decibels, another familiar measurement. If you wish to visualise amplitude, you should think of the height of the waveform. The higher and deeper the peaks and troughs are, the louder they will be. The shorter and shallower they are, the quieter they

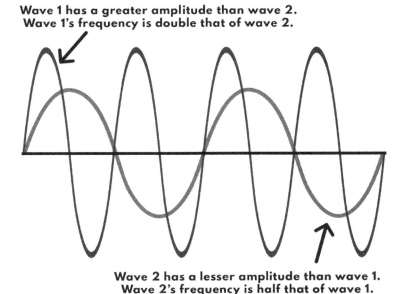

Figure 9.1 Two sinewaves illustrating contrasting frequencies and amplitudes.
Source: Image by Sam George.

will be. Again, you are most likely to instruct your oscillator of this information without thinking about it. The velocity with which you play your MIDI keyboard will probably instruct the oscillator of your desired amplitude.

So far, so good then. No nasty surprises here. And as we continue through this topic, you'll find that there aren't any nasty surprises at all. Everything is entirely logical.

> TASK – Look back at your range of synths. Identify where the oscillators are. Some synths may only have one; others could have two, three, four, or more!

Day 4 – Waveshapes

The third way of manipulating an oscillator is through the selection of waveshape. The good news here is that the primary options you have at your disposal are limited to just four individual shapes. These shapes are sine, sawtooth, square, and triangle, and each has its own identifiable characteristics.

Before we dive into these characteristics, understand this: Musical sounds are generally not just made up of one frequency but are a combination of multiple different frequencies, called overtones or partial tones. The lowest frequency (the fundamental) is what we perceive to be the sound's pitch. All the other partial tones combine to create the sound's unique timbre. Bearing that in mind, let's define our waveshapes.[2]

A sine wave is clean and smooth. It is as basic as sound can get. A sine wave has no overtones. It is a pure fundamental. The sound you make when whistling is about as close as you can humanly get to creating a pure sine wave. Sine waves are great for making deep, smooth sub-bass that doesn't interfere with other mix elements.

Square waves are rich and buzzy. In addition to its fundamental, a square wave will have multiple harmonics. To clarify, an overtone is any higher frequency standing wave, whereas harmonics are integral multiples of the fundamental. In a square wave, the harmonics occur in whole odd-number multiples of the fundamental. Combined with the fundamental, these harmonics give the wave its square shape. They can make crunchy, aggressive kick drum sounds.

Triangle waves have the same odd harmonics as square waves. However, rather than consistently disappearing into oblivion, in a triangle wave, they taper off, providing the triangular shape. Its sound is somewhere between a sine and a square wave. It's not as smooth as a sine but not as buzzy as a square. It's clearer and brighter than a sine wave. Triangle waves are often likened to reed instruments such as recorders and flutes and are great for lead lines.

Sawtooth waves are jagged. They're the buzziest-sounding and are even harsher-sounding than square waves. This is because they're the richest in terms of harmonics. So, if you're looking for something in-your-face, a sawtooth is what you want.

If you're feeling wholly bamboozled with technical jargon at this point, don't panic. For our purposes, none of it matters. All you need to remember is that you have four primary wave shapes, and they each have their own personality.

I stated the 'primary options' previously because stemming from these four shapes come all manner of variations. But ultimately, they all have their roots in one or other of these origins.[3]

TASK – Set up a blank patch in your simplest synth. Ideally, use a synth with just one oscillator. Listen to each wave shape so you can begin to learn their sounds.

Day 5 – Combining oscillators

Individual oscillators on their own won't get many people excited. Ultimately, a single wave shape in isolation will not inspire anyone. The sound from a single oscillator will be comparatively thin and static. The magic happens when you combine two or more oscillators. There are limitless ways to combine different wave shapes from different oscillators.

But why would you want to? Let's take an orchestra as an example. A solo violin playing a melody on its own is pleasant but nothing unusual. A solo violin playing a melody in unison with an oboe and a trombone? Now that's thought-provoking. This concept is effectively what you're doing with your oscillators. You're combining different timbres creatively to create a unique sound.

Most synths will have at least two oscillators, and it's common to have up to four. This gives you a massive array of possibilities with which to experiment. If your synth allows you to adjust each oscillator's output levels and pan positions individually, you can begin to get creative. But the fun is only just beginning.

TASK – Explore combining different waveshapes from multiple oscillators. Listen to how the complexity of the sound grows as you add more oscillators of different shapes.

Day 6 – Tuning, unison, and voices

On each oscillator, you will find that you can adjust the pitch. This pitch adjustment will most likely be in three stages: First, by octave. You can pitch oscillators up or down in whole octave increments. This is beneficial because you can create a sound across a wide pitch range. The second stage is by step or semitone. Tuning oscillators up or down in semitone increments allows you to create harmony between them. For example, tuning one up by seven steps will make an interval of a perfect fifth. The third stage is in cents. There are 100 cents to every semitone, so this is effectively your fine-tuning. The immediate question that arises here is, 'Why would you want to make your oscillators out of tune with each other?' Good question. To answer it, let's again consider our orchestra. Each instrumental section in an orchestra contains multiple players. Orchestral musicians tend to be highly accomplished, but even so, the chance of them all playing 100% in tune is zero. There will always be subtle variations in tuning between different sections or players, particularly as their vibratos will move at different times. Raising or lowering your oscillators in cents is a step towards recreating this effect.

You'll notice I just mentioned multiple players performing each part. But currently, we only have a single representation of each oscillator. This is effectively the same as having just one player on each part. The solution here is to increase the number of players. Some synths will allow you to do this on a per-oscillator basis, with a unison control. Increasing the unison will create an additional version of that oscillator that is slightly detuned from the original. More advanced synths will allow you to dictate precise details about how these additional layers are detuned. By increasing the unison of an oscillator, you are effectively adding players to the part, stacking one player on top of another. The oscillator gets thicker and richer as the layers increase, with the overall timbre complexifying.

Now is a good time to mention that your synth will have a global control entitled 'voices'. This control allows you to specify the maximum number of notes your synth can produce at any time. If you only have one voice, you can only play a single note at a time, effectively making it monophonic. Increasing the voices will enable you to increase the polyphony within the parts you play.

TASK – Investigate the oscillator(s) on your synth(s). Identify how to adjust the tuning. Can you do it per oscillator, or is it a global setting? Find the voices control too. Again, is this per oscillator, global, or both?

Day 7 – Filters

As a quick recap, the sound created by your oscillator is likely to contain a fundamental and a harmonic series. These elements will combine uniquely and can be defined as the instrument's timbre. Timbre is commonly described adjectively, with words like warm, gritty, and silky.

After your oscillator(s), the sound will travel to a filter section where you can shape its harmonic characteristics. You'll encounter four options: Low-pass, high-pass, band-pass, and notch. These all work in the same way as on your equaliser, but let's refresh our memories for the sake of completeness.

A low-pass filter will set a cutoff frequency allowing everything below it to pass and rejecting everything above it. This filter type is commonly used for dark and warm sounds where you want to restrain your sound.

High-pass filters are the opposite. They set a cutoff frequency that allows everything above it to pass and cuts everything below it. They're primarily used to remove unwanted low frequencies. They're good for crisp and bright sounds.

Band-pass filters allow you to select a group of frequencies that you want to allow through, cutting the rest. This technique is often used to emulate formant frequencies of the voice and is good for nasal sounds.

Notch filters are the opposite of band-pass filters. They allow you to specify a group of frequencies that you wish to prevent from passing through. They're commonly used to remove specific, unwanted frequencies, but have other creative applications, too.

Using filters is the most basic way there is of shaping a waveform. However, they have one extra secret weapon that has the power to make things a whole lot spicier: Resonance. The resonance control will boost the frequencies around the cutoff point. This will create a ringing sound which is ideal for making the cutoff frequency point more recognisable. Proceed with caution, though, as too much resonance will become piercingly irritating. The resonance boost is integral to the classic filter sweep sound, which we will explore further later.

> TASK – Identify the filters section on your synth. How does it affect the oscillators? Can you assign it to specific oscillators, or is it a global filter?

Day 8 – Amplifiers and envelopes

After your signal has been filtered, it reaches the amplifier section of the synth. Every amplifier will be controlled by an envelope. This envelope dictates how the volume of your synth changes over time and commonly uses four stages known as ADSR. Understanding the ADSR envelope fully is key to understanding how to shape your sound.

The A stands for attack. The attack denotes the length of time it takes the signal to reach its peak level after playing your MIDI keyboard. Short attack times mean your sound will reach full blast quickly, whilst long attack times mean your sound will gradually fade in.

The D stands for decay. The decay indicates how long it takes for your sound to fall to its sustained level once it has reached its peak.

The S stands for sustain. The sustain is the level at which the sound will hold after it has risen to its peak level and decayed down and is measured in decibels. Note that a sound will only sustain if you hold a key on your keyboard. If you simply press and release, the sustain step will be missed.

The R stands for release. Once you stop holding your MIDI note, the release dictates the length of time it takes for the sound to return to being off. Short times mean the sound cuts off swiftly, whereas longer times mean the sound will fall away gently.

On some synths, you will come across a fifth step in this chain. H stands for hold and comes after the attack. The hold will denote how long the sound will remain at its peak level before decaying down to the sustain level.

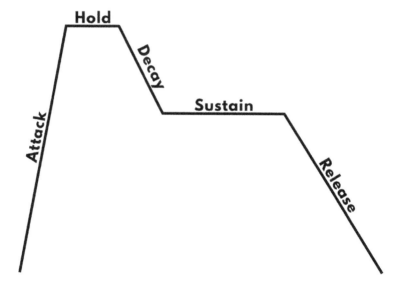

Figure 9.2 An AHDSR envelope.

Source: Image by Sam George.

Understanding A(H)DSR envelopes is vital because you don't just use them to control a sound's volume. You can use them to affect any other parameter within your synth too.

> TASK – Spend time getting to know the A(H)DSR envelope. Ensure you understand exactly what each stage in the process does and how it affects the sound. Look at amp envelopes on different synths if you have more than one. Are they presented differently?

Day 9 – Modulation

Modulation is where synthesis gets interesting as it's the stage where you start to create movement in your sounds. If you substitute the term modulation for movement, it's much easier to comprehend.

Using an envelope is the most common way to create movement in your sound. You can assign an envelope to almost any parameter within your synth to make it move once in a specific way. The simple principles of envelopes that you just learned apply the same way: You can specify the A(H)DSR of any parameter. The trickiest concept to grasp is that, rather than the envelope controlling the volume, it controls the position of something else.

Let's look at a couple of examples to clarify this: The most common parameter to modulate with an envelope is the cutoff frequency on the filter. Picture this: The attack quickly modulates the cutoff frequency from around 500Hz to 5kHz. The decay quickly brings the cutoff down to 1kHz, where it sustains for a second before slowly releasing back to its original position at 500Hz.

Let's consider another example. Imagine the pitch of your oscillator has an envelope assigned to it. The attack tells the oscillator to rise quickly by an octave. There is no decay. The sustain is at full, so the pitch stays an octave up whilst you hold the key down, and the release quickly falls back down an octave to the original pitch.

There are two other important things to be aware of regarding your envelopes. First, there are different trigger types. You may be able to tell your envelope to re-trigger only after all current MIDI notes have been released. This is commonly called 'single' mode. 'Multi' means the envelope is re-triggered by every MIDI note. These are the most common, but there are others too.

Secondly, you will be able to assign one envelope to multiple parameters. You don't need to set up a new envelope for every modulation assignment unless you desire a different shape.

The unique factor to be aware of with envelope modulation is that it triggers the movement once only. This is the opposite of an LFO.

> TASK – Set up a modulation envelope and assign it to different parameters in your synth. Try it on your oscillator's pitch, panning, and voices. Adjust this envelope's parameters to see how it affects the sound.

Day 10 – LFOs

LFO stands for Low-Frequency Oscillator. Technically, it is another oscillator in your synth. However, its frequency is so low that you can't hear it. LFOs can take any waveshape (dependent upon the limitations of your synth) just like a regular oscillator. Their unique characteristic is that they are continuously cycling around their waveshape rather than modulating a sound once only as with envelopes.

You can adjust your LFO's cycle rate in one of two ways: Either by frequency (in Hz) or by tempo-syncing it to your project and using duration values such as one-sixteenth, one-eighth, and so on.

So, what are some common things to modulate with an LFO? The panning position is one, where you want the sound to move from left to right and back again continuously. Tuning, specifically the cents of an oscillator, is another where you want the pitch to fluctuate constantly, creating a kind of vibrato sound. You could modulate your filter's cutoff frequency with an LFO so that its position moves up and down continuously. Or what about your oscillator's volume to create a tremolo effect?

The keyword to associate with LFOs is continuous. LFOs don't stop after one trigger; they cycle around and around.

The standard place to start with LFOs is with a sine wave. Due to their smooth nature, an LFO with a sine waveshape will give you a smooth modulation effect. The more interesting your LFO waveshape, the more interesting the resulting modulation effect.

> TASK – Set up an LFO and assign it to the same parameters you assigned your envelope to yesterday (ensure you remove the envelope parameters). Experiment with tempo-synced and free cycle rates. Explore different LFO waveshapes.

Day 11 – Other possible features

We've looked at all the most likely features you'll encounter in a synthesiser. However, there are many other things that you may come across that will make that specific model unique. Whilst this list is by no means exhaustive, it will cover most of the bases.

Sub oscillators are common. Typically, they are linked to the principal/first oscillator and are pitched an octave lower. They are most commonly a sine wave whose job is to thicken and bolster the overall sound without contributing additional harmonic richness.

You may well also come across a noise generator. This could be part of one of your main oscillators or may run through its own independent oscillator. Commonly they will offer white and pink noise and are there to provide textural complexity to your sound.

You'll frequently encounter equalisers within your synth. These tend to be global, i.e., affecting the whole synth rather than individual oscillators (as you already have your oscillator filters to play with), and usually will be limited to three or four bands rather than giving you complete parametric control.

Most modern synths will have an effects section included. In some synths, the options here are exhaustive. In one of my favourites – Ana 2 from Sonic Academy – you have a wide selection of reverbs and delays, all manner of different distortion types, multiple modulation options, a whole load of dynamic processors, and some other hidden gems, too! You'll find similar ranges in other major players such as Massive and Serum.

The next things you could come across are MIDI effects. Your synth may include an arpeggiator or even a chord trigger section. Sometimes, you'll even come across a step sequencer, which you can use as an additional modulation option instead of an envelope or LFO.

Continuing along the alternative modulation route, you may see an MSEG, which stands for Multi-Stage Envelope Generator. Think of this as an envelope, but instead of having the fixed A(H)DSR points, you can draw in as many points as you like to create a unique modulation shape. You can then use it as an envelope to trigger once or link it to an LFO to continuously cycle.

Finally, on this list is a modulation matrix. On more complex synths, a mod matrix is key to controlling what goes where. Put simply, a mod matrix is the summary and control panel for all your routing.

As you may have realised by now, synths can become your one-stop shop for all your sound design needs. Creating your unique sound and applying EQ, compression, and effects within the same plugin can be great not only for keeping your session tidy and clutter-free but also for preserving some vital CPU.

TASK – This is the crucial step. Learning all of the individual quirks of your synths will allow you to get the most out of them. Learn what every section of your synth does. Know it inside out. If you have more than one, learn one thoroughly before moving on to another.

Day 12 – What makes a synth unique?

With the basic principles of synthesis being the same regardless of make or model (both in the analogue and digital domains), you may ask, 'What makes one synth different from the next?' This is a very fair question.

The starting point in the answer is the number of oscillators a synth has and the waveshapes they support. Whether or not you are able to have one oscillator to modulate the frequency of another is part of the consideration. But there are several other factors.

The number of filters, their types, and how you can route them determine how you can shape your sound. The filter modes they support (high-pass, low-pass, etc.), their characteristics (slope, resonance, etc.), and whether you can arrange multiple filters in series or parallel all contribute to a synth's uniqueness.

The quantity and type of envelopes and LFOs you may have also matter, as well as the presence or absence of a noise generator. Whether the synth has a fixed routing architecture or a free modulation matrix is also crucial. For example, do LFO1 and Envelope1 always affect Oscillator1, or can they be assigned to another oscillator?

This all mattered a lot in the analogue domain, where the physical construct of the synth denoted its limitations. This is still the case in digital synths that are designed to behave like analogue ones. However, in many modern synths constructed to have versatile workflows, there isn't much difference between them. Most modern synths are used because they support sample playback. If you're using a sample as your starting point, this can easily be replicated in another synth.

However, there's one massive subject that I've overlooked until now that justifies the existence of such a range of synths. This is the synthesis type.

> TASK – Examine two or more different synths. Write a list of what similarities and differences they have. Do you have a preference between them?

Day 13 – 1/10: Subtractive synthesis

There are arguably ten different types of synthesis. Some will say there are more or less, but generally, ten is the agreed-upon number. We'll cover each of their features and differences now.

The first type of synthesis, which is also the most common, is subtractive synthesis. In this case, you start with something that is harmonically rich (your oscillator(s)) and then remove parts of its harmonic content with a filter and shape its volume with an envelope. Think of it like sculpting a block of stone. You start with something significant and chip away at it until you're left with everything you want.

This synthesis style is generally considered an analogue method, but it is often replicated digitally through analogue modelling. Classic models include the Moog Minimoog Model-D, Arturia Minibrute, and Roland Jupiter-8.

Some know subtractive synthesis as East Coast synthesis as it was created by Bob Moog, who was based in New York.

> TASK – How you approach the task for all ten synthesis types will be up to you. You may have that type of synth already at your disposal. In which case you can explore it. It may sound particularly interesting, in which case you may wish to get one to play with. Or you may just want to watch a few YouTube videos to get a better feel for what I'm talking about. The choice is up to you.

Day 14 – 2/10: FM synthesis

Next on our list is FM synthesis. The FM stands for Frequency Modulation. This is a digital form of synthesis where one waveform modulates another. Analog FM synthesis does exist, but typically when FM synthesis is referred to, people mean the digital kind. More specifically, they probably mean the form of linear FM synthesis used in the Yamaha DX7 and other similar models.

FM synthesis begins with a pure sine wave which is called a carrier. A second inaudible sine wave called the modulator then modulates the carrier. This modulation creates harmonics in the original carrier wave. You can combine multiple carriers and modulators to create more complex sounds called algorithms.

Yamaha patented FM synthesis, meaning no one else could use it. But other companies have subsequently come up with their own variations on the theme. PD (phase distortion) is Casio's version of FM. VPM (variable phase modulation) is Korg's version.

Day 15 – 3/10: Sample-based synthesis

Third on the list is sample-based synthesis. Although this is commonplace now, this concept was novel and expensive in the late '70s and early '80s. Some of the first sample-based synths, such as the New England Digital Synclavier and Fairlight CMI, were eye-wateringly expensive.

Sample-based synthesis can be accomplished on a few different types of machines. The most obvious is a sampler. A sampler lets you take an analogue sound and transform it into the digital realm, enabling it to be manipulated and played back. There are plenty of affordable hardware samplers these days, but your DAW will almost certainly come with at least one built in.

ROMplers will offer similar sound-design features, but the main difference is that you can only use the preloaded samples that come on them. Lots of drum machines fit into this category. The infamous Kontakt from Native Instruments is an excellent software example of a ROMpler. These synth styles put vast amounts of readymade sounds at your fingertips, with minimal tweaking required to achieve the desired sound.

Hybrid synths sit somewhere between a digitally sampled waveform and analogue subtractive synthesis. This means you'll have a digital oscillator that contains the sample that you can then affect with filters and envelopes.

Deriving from this, 'Sample and Synthesis' is a broad term covering a kind of sample-based digital synthesis that rose to fame in the late '80s with the Roland D-50's LA (Linear Arithmetic) synthesis. Here, you've got a very short transient sample paired with a subtractive-style oscillator.

Day 16 – 4/10: Wavetable synthesis

Wavetable synthesis has become wildly popular recently, but it originates in the late '70s. It is known for being highly flexible and for being excellent at creating evolving sounds. Its core comprises digital oscillators that use wavetables or groups of single-cycle waveforms.

The playback can move laterally across all the contained waveshapes, which generates a unique feeling of movement and sonic progression. As in subtractive synthesis, you can control this movement, or modulation, with envelopes.

Palm's PPG was the first successful hardware wavetable synth. It paved the way for modern-day staples such as Xfer Records' Serum, Massive X from Native Instruments and Pigments by Arturia.

Day 17 – 5/10: Vector synthesis

Vector synthesis is a bit of a nonstarter in the synthesis world. It still appears occasionally in modern synths, but overall, it's obsolete.

Imagine a square. In each corner of the square, you have a different sampled waveform. With a joystick or XY pad you can adjust the balance between the four different waveforms. This isn't so different to wavetable synthesis, where you're laterally moving between waves that are stacked side by side. But for whatever reason, vector synthesis never really took off.

With vector synthesis, as with wavetable synthesis, you can adjust the position of the XY blend with envelopes and LFOs. This invites some pretty exciting sound combinations when you consider that you could be moving between four entirely different waveshapes in a short period.

Some stand-out models of vector synthesisers are the Prophet VS, Yamaha SY, and Korg Wavestation.

Day 18 – 6/10: Additive synthesis

In contrast to subtractive synthesis, which begins with a harmonically rich sound, additive synthesis is the opposite. You start with several small pieces of sine waves that you add together to create something that is harmonically rich. A great example of this is a Hammond organ. On a Hammond, you have loads of drawbars that add harmonic content to the sound as you draw them in.

The best modern example of additive synthesis is Native Instruments' Razor. This synth demonstrates the ability to create almost any waveform using solely sine waves. Additive synthesis' main benefit is its ability to be precise and well defined.

Resynthesis is a process that is strongly associated with additive synthesis. This occurs where a sound is analysed and recreated using sinusoidal harmonic partials (which simply means partial sine waves). Rather than setting out with a sine wave and trying to build a new waveform from scratch, the process of resynthesis begins with the waveform and attempts to reproduce it. However, this is not a particularly common form of synthesis, so you're unlikely to encounter it in everyday usage.

Day 19 – 7/10: Spectral synthesis

Spectral synthesis is a sort of resynthesis. We're into the realms of pretty niche synthesis here. A sound is transformed into several 'bins' where each one represents the original sound's frequencies. This process is known technically as multiresolution sinusoidal modelling. In simpler terms, a custom filter bank analyses peaks and other elements within the frequency spectrum of the original waveform. Some other stuff happens that I don't need to trouble you with here, and you end up with a graphic representation of your sound on a spectrogram.

The key feature here is that you can then draw and paint on this graphic to alter how the waveshape sounds. This is fun and is an excellent way of creating complex sounds that would be almost impossible to program using a different type of synth. If you want to play with spectral synthesis, I strongly recommend iZotope's Iris plugin, or if you're a Logic Pro user, then you have it built within Alchemy too.

Day 20 – 8/10: Physical modelling

I mentioned much earlier in this unit how you could use subtractive synthesis to mimic the sound of real instruments. Physical modelling goes one step further and genuinely attempts to recreate a sound by digitally emulating the processes that make up a sound. Physical modelling uses digital signal processing (DSP) to reproduce different elements of the sound. This usually is broken down into three parts: The exciter (such as the breath in a wind instrument or the bow on a string instrument), the instrument's resonant body, and the material the instrument is made of (metal, wood, etc.).

There are a couple of different types of physical modelling synths. Karplus-Strong string synthesis is the first. This loops a short waveform through a filtered delay line. Modal synthesis and resonator synthesis recreate the sound of real instruments by using band-pass filters.

The world of physical modelling is a complex one. The Yamaha VL1 and Korg Prophecy are good starting points, or Sculpture that comes with Logic Pro is an excellent soft-synth option.

Day 21 – 9/10: Granular synthesis

As its name suggests, granular synthesis works on a much smaller scale than anything we've seen so far. It's a sample-based synthesis that breaks the sample down into tiny parts or grains and allows you to jump around the different grains during playback. As well as playing back the grains out of order, you can also play multiple grains simultaneously, which allows you to create some interesting 'clouds.' You can manipulate the grains' size, volume, position, and so on to create new sounds. Another interesting feature here is how you transition between the grains. You'll want to put fades on the grains to avoid clicks and pops. How you fade is called the 'window', and it will affect the overall timbre significantly. The Tasty Chips GR-1 is a good example here.

The sounds and textures you can come up with using granular synthesis tend to be quite unworldly. For this reason, sound designers often use it in video games, movies, and TV. As it's a popular method of sound design these days, it can be found in most DAWs: Ableton's Granulator II, Logic Pro's Alchemy, Bitwig's Sampler, Reason's Grain Sample Manipulator, as well as premium options like Straylight from Native Instruments.

Day 22 – 10/10: West Coast synthesis

Last on the list, and to bring this tour full circle, is West Coast synthesis. Concurrently with Bob Moog doing his thing on the East Coast, Don Buchla was doing his in San Francisco. Whilst East Coast subtractive synthesis begins with a complex waveform, West Coast commences with something relatively simple, such as a triangle wave. It then utilises waveshapers to add harmonic content. The filtering is done through a low-pass gate, which performs the role of both a filter and an amplifier at the same time.

West Coast synthesis is less focused on musical applications and more on experimentation. For this reason, you'll tend not to find the traditional piano keyboard attached. You're more likely to encounter sequencers, touch panels, and other devices that allow you to explore the sound in a non-chromatic fashion.

An original Buchla Music Easel is hard to find, but the Buchla Easel V emulation from Arturia does a great job.

The reason I've spent time explaining these different forms of synthesis is not so that you can become an expert on them all. But by being armed with this basic information, you should now be able to make better decisions about what synthesiser you want to reach for within your DAW. What is the best soft synth for the job? Is it the wavetable of Serum, the granular of Straylight, the spectrum of Iris, or the ROMpler of Kontakt? By choosing the proper synth for the sound you're looking for, you'll be presented with a range of presets more suited to the purpose. And in turn, you'll then be able to tweak those presets more quickly to achieve the desired tone.

Day 23 – The main synth sounds

Several synth sound categories denote the sort of sound that you are creating. If someone asks you for a pad, you should know what this means and not give them a lead sound. You shouldn't provide them with a pluck if they want a bell. Let's cover each primary sound category now, outlining the key features of each.

- Leads: These are generally monophonic, meaning they can only play one note at a time. They are bold, with rich harmonic content and strong sustain, which helps them cut through the mix.
- Bells: These have fast attack and decay times but an extended, prolonged release. Again, they tend to be harmonically rich and are often made using FM synthesis.
- Pads: These have long attack and release times and a high sustain level. They frequently emulate the sound of a choir or a bowed instrument.
- Keys: These will always be polyphonic, meaning they can play more than one note at a time and will emulate the sound of a piano or organ. They tend to use simpler waveforms and are less harmonically complex. This allows chords to sound more coherent.
- Plucks: These have a fast attack, decay, and release and emulate the sound of a pizzicato (plucked) string instrument or palm-muted guitar. They are like bells but are less harmonically rich and have shorter release times. They often utilise Karplus-Strong synthesis to create a more convincing attack.
- Brass: These have a slightly slower attack and a fast release. They are harmonically complex and extend over a wide frequency range. Like bells, they're often made using FM synthesis.

- Bass: These are almost always monophonic. They tend to be built upon a sine wave, which ensures a solid fundamental frequency and will often have another waveshape layered on top to add harmonic complexity.

Almost all synth sounds you can create, on pretty much any synthesiser type, will fall into one of these categories. By exploring each of them, getting to know them, and learning how to dial them in with pace and accuracy, you will speed up your workflow and productivity to no end.

> TASK – Practise creating each of these synth sounds and learn to identify its characteristics. You could try picking presets randomly and naming their type as a practice method.

Day 24 – 20 top tips: 1–5

Now that you're armed with all the essential information on synths, it would be easy for me to wish you well and send you on your way. However, I'm all too aware from personal experience that the world of synthesis is a vast and daunting one. My general advice is to focus on the fundamental areas first, ensuring you understand oscillators, waveshapes, filters, modulation, and LFOs correctly. Focus on subtractive synthesis, which is the most common and easiest to comprehend. Then, rather than trying to conquer all nine other synthesis types, pick one or two to explore. There are many synth-nerds out there who know a lot about the subject, and even the most knowledgeable of synth gurus won't claim to be an expert in all ten types.

What I think will be helpful for you at this stage is to give you some top tips to help keep you creative, help keep you in the flow and prevent you from getting bogged down. So, here are the first five:

1. Not all parameters are created equally. Some controls will affect your sound much more significantly than others. If you're looking for significant changes, reach for your filters, amp attack and release, and LFO depth. And if you want to fatten or warm up your sound quickly, look to detune your oscillators.
2. The quickest way to access new tones is by adjusting your filter settings. Get creative with what is modulating your cutoff frequency. You can try modulating it with keyboard velocity or tracking, envelope amount, and even the filter envelope settings. Another simple option for quick results is to try some different wave shapes. Look to explore shapes beyond the primary ones. The more complex the oscillator wave shape, the more complex your sound will become.

3. Prior planning is vital, especially when you're under pressure. If you're in a session and need to dial in a sound quickly, then turn to a synth that you know. Keep your patches well organised within the synth's library, including any custom patches you've created beyond the stock. In short, don't let your lack of knowledge stifle your creativity. Allow your limitations to be a positive. Utilising a limited palette to paint with is not a weakness.
4. Conversely, go wild when you're in less of a hurry and have time to experiment! A great technique to explore is layering sounds together. If you select patches from the same synth to layer, they're likely to blend well as they come from the same sonic location. You can broaden this out and layer patches from similar synths, such as a Roland Juno and JX.
5. Arrangement is a term used to decide which instrument or sound will perform a particular line or melody within a song. Orchestration is another word for this. Your synth sounds are part of your arrangement. When considering what sort of synth sound you want, think in terms of time and tone. Notes that last longer affect your arrangement differently than shorter ones. Pads and evolving sounds can quickly fill holes in an arrangement, adding depth and space.

> TASK – Each of these tips contains information you can take and apply to your practice. Today, and over the next three days, explore each of these tips systematically.

Day 25 – 20 top tips: 6–10

6. Tempo is everything. The timing of moving parameters (such as LFOs) often needs to be in time with your track. This prevents your track from becoming too messy. Ensure you know where the sync option is for these parameters so you can link the modulation to your project BPM. You can then experiment with automating rates. For example, if your LFO is set to a half note, try increasing the rate to a quarter note as your track builds to increase the rhythmic tension. While talking about rate, vibrato is often linked to tempo, although traditionally, its rate is much freer. So, try automating your vibrato without it being linked to your project's tempo.
7. This one is more complex, but it involves recording performances with real-time edits. Consider a step sequencer into which you make various edits and parameter changes as part of a performance. It may appear at first glance that not all the changes you made during your performance were captured. In this instance, see if you can find your synth's transmit/receive parameters. You'll probably find them in the global or MIDI settings of the synth. Often, synths that appear simple on the face of it can

be deeper than you think. Ultimately, this comes down to knowing your instrument.
8. Adjusting parameters whilst a synth is sounding can add a load of expression and dynamism to a performance. Some controller options to play with are your pitch and mod wheels and joysticks, knobs and rotary controllers, keyboard velocity and aftertouch, and foot pedals. You can try linking any of these controllers to pan, pitch, volume, cutoff, and LFO amount for quick results. Many presets will already have controller parameters assigned to valuable things within the instrument.
9. Use your sequencer as a control module to animate your sound. Don't feel that you need to record notes and controller changes simultaneously. It's very common to perform notes first and controller changes afterwards. This lets you focus on playing the notes correctly first and then adding the creative movement in a second performance thereafter.
10. The most important part of a sound is usually the initial attack. It provides a lot of information to the listener about the sound. If your attack doesn't fit in with the overall style of the track, then it can be misleading or confusing. So, pay careful attention to the front end of your sounds, especially if you've changed tempo or if your tempo changes during the track, as you may need to adjust your attack to sit better at different tempos.

Day 26 – 20 top tips: 11–15

11. Try not to be limited by the same handful of modulation options. Try to keep your thinking fresh and explore different combinations. Have you tried modulating the resonance on a low-pass filter? What about adding a slow LFO onto a cutoff filter? Try modulating your pan position with keyboard tracking. Or use your mod wheel to affect various envelope parameters. Modulate filter parameters by velocity. There are limitless routes to explore. Don't be bogged down with the same handful.
12. Put effort into exploring all areas of your synth. Don't be satisfied with just knowing the basics of it. Be hungry to learn it properly. This is good advice with any plugin! Synths often have more depth than may first meet the eye. Investigate each menu, listen to all the effects, and try out all the different wave shapes. Be curious.
13. Here's a tip for a great bass patch: Rather than having your sine wave going through the filter so that it's affected, ensure the sine's oscillator isn't routed through the filter. This will give you the constant low end. If you can't assign your oscillator routing individually, try layering two patches to create the same effect.
14. If recreating vintage sounds from classic synths becomes your thing, explore the features of the original hardware models that made them unique. Most analogue synths have their individual quirks. The Minimoog isn't

touch-sensitive, for example. It has three oscillators, but if you want an LFO, you can only use two of them. Understanding these things will limit how you can route and modulate things, forcing your hand to move creatively in other directions.
15. Sometimes, no matter how hard you try, you won't be able to recreate a sound satisfactorily. Sometimes, you just need the real thing. Certain types of sounds are invariably associated with their creators. Sometimes, if you want a specific sound, you may just need to bite the bullet and purchase an emulation of that synth. Almost all synths are modelled somewhere by someone. The places I look first are IK Multimedia in their Syntronik 2 and Arturia in their Analog Lab V. Both have a massive range of vintage synth emulations.

Day 27 – 20 top tips: 16–20

16. To access a whole new palette of new sounds quickly, turn off any onboard effects you're running within the patch, and add your own effects after the synth. Distortion is used a lot nowadays to create more robust, aggressive sounds. It's worth playing with as much as delay, reverb, chorus, etc.
17. Often less is more. It's easy to get carried away, piling layer upon layer to create a massive wall of sound, only to find that it doesn't provide the desired punch in the mix. More layers can often mean more mush and less precision. If you want to increase the punch and clarity, peel back some layers.
18. If you feel as though you're dialling in patches on autopilot, it may be time to switch up the roles. If you unconsciously reach for Native Instrument's Monarch for a synth bass sound, try hitting reset and using it for your pads instead. If Omnisphere is your go-to for pads, use it for your bass. You may not necessarily create the desired result immediately, but it will force you to continue to learn and develop. And developing lateral thinking and problem-solving skills is vital in enabling you to get out of a hole when you least expect it.
19. Don't rush out and blow loads of cash on a library of different synth models. Just as with plugins, you should spend time learning what you have. There's nothing wrong with using whatever comes as stock with your DAW. And if you want to buy something, buy one synth, and learn it inside out, upside down. Exploring a synth thoroughly will usually proffer better results than superficially looking at three or four models.
20. Lastly, don't be a snob about presets. Using a preset as a jumping-off point is synonymous with the modern producer's workflow. It's not cheating. The critical point is to use it as the starting point, not the end. Mould it, sculpt it, make it your own.

Day 28 – Unit summary

As I said at the outset of this unit, the goal here was not to turn you into an expert in synthesis. The goal throughout this whole book is not to turn you into an expert in anything. Like any of the other chapters you've already read, synthesis is a skill that can be developed over time, like EQ, compression, or anything else. Don't allow it to intimidate you. It's one of the most impressive skills you can have in your arsenal. Just imagine the look on your partner's face when they say, 'I'm imagining a phat synth bass sound like that Justice track, you know?' and you can reproduce something similar within a couple of minutes. Now that's impressive! You'll never get the same instant gratification by compressing something well or applying some superb dynamic EQ. Nothing else has the same wow factor.

Let's recap. In this unit, we've covered:

- All the essential features of a synth, including oscillators, waveshapes, tuning, unison and voices, filters, envelopes, modulation, and LFOs
- Different synthesis types: Subtractive, FM, sample-based, wavetable, vector, additive, spectral, physical modelling, granular, and West Coast
- The different synth sounds: Leads, bells, pads, keys, plucks, brass, and bass
- And 20 top tips to keep you in your creative flow

I hope you're feeling inspired to throw yourself headfirst into the subject. Don't let it scare you. Own it!

I've decided not to include a checklist at the end of this chapter. Whilst synthesis is a wonderful skill to possess, it isn't as fundamental as everything we've covered so far. So, I'll understand if you wish to proceed without conquering it all!

Further reading

1 Burry, M. (2021). *How we hear: A step-by-step explanation.* [online] healthyhearing.com. Available at www.healthyhearing.com/report/53241-How-we-hear-explainer-hearing [Accessed 9 Nov. 2022].
2 Hopkin, B. (2022). *Fundamental, harmonics, overtones, partials, modes.* [online] barthopkin.com. Available at https://barthopkin.com/fundamental-harmonics-overtones-partials-modes/ [Accessed 9 Nov. 2022].
3 Aulart. (2022). *Oscillator waveforms: Types and uses – part I.* [online] aulart.com. Available at www.aulart.com/blog/oscillator-waveforms-types-and-uses-part-i/ [Accessed 9 Nov. 2022].

Unit 10

Mastering

Day 1 – The dark art

How often have you heard that mastering is a dark art that only a few blessed individuals are competent enough to undertake? To be in this crowd, you must have graduated from Hogwarts or been trained by Yoda himself. And even then, the journey to understanding mastering entirely is an uncertain one that only the foolhardy would embark upon.

This is a lie that is spun primarily by established practitioners motivated by self-interest that try to justify their existence by bamboozling you with terms like metadata, dithering, and LUFS to put you off challenging the subject yourself. But, like anything, mastering is a skill that you can learn. You don't need to receive a divine gift. You can train to become competent at mastering, just as you can to balance a mix, play backgammon, paint still life pictures, or anything else.

Mastering comes with its own specific set of processes that you must follow. But I believe that most of its challenges lie in its technical requirements and limitations, most of which you don't even need to know by heart. Having them written in your notepad next to you is fine.

In this unit, I will do my best to draw back the curtain and lift the veil on the mystique of mastering. I'll explain the steps you need to follow in simple terms, avoiding any overly complex language. When I'm done, I hope you'll feel confident enough to tackle the subject with enthusiasm rather than scepticism. So, let's begin.

Day 2 – Mixing vs mastering

Often, I hear people struggling with the difference between mixing and mastering. Where does one end and the other begin?

To understand this clearly, I like to use the analogy of a sports team. Let's use soccer as our example. Mixing is like the coaching and training of individual players. Players are trained to perform their best in certain situations and to be able to respond to specific pressures in ways that are appropriate to them. They train in ways that are specific to their position on the field. The way a defender trains is different to how a goalkeeper prepares. This individual training is like mixing individual components in your mix, your kick, bass, lead vocal, etc. Your method for mixing individual parts is unique to that part.

You may also train groups of players together. The whole attacking unit may prepare together, for example. This will help with cohesive movement and coordination of the players belonging to that section. This is like your buss processing. You can process or train groups of instruments together. For example, you may process all your drums together or all your vocals.

Therefore, mixing can be considered as the micromanagement of the channels within your track. It focuses on the individuals, on the small details that allow each cog in the machine to operate effectively. Without this micromanagement, you would have an uncoordinated, incohesive machine.

Mastering is the process of managing all the individual components together. It's like the manager of the team. You can attempt to manage a team without first coaching it, and it will undoubtedly help, but it will be much less effective than managing a well-coached unit. Think of it as the team's tactics. It's the overall tactical decisions taken from one game to the next that are adapted depending on the selected team and the playing environment they are entering. In mastering terms, it's the decisions taken about the entire track. These decisions affect everything, not just individual components. Therefore, these decisions must be taken with consideration of the bigger picture. That means ensuring that the decisions you make are beneficial for the whole track, not just one or two individual elements within the track.

To put it much more simply, mixing is the hard work. It's the nuts and bolts, the bread and butter. Mastering is the polish and shine that comes on top.

> TASK – Consider tracks you've previously thought of as ready for mastering. Given what we've just discussed, are further mixing steps required to prepare them fully for mastering?

Day 3 – A brief history

Mass-produced music became commonplace in the late 1950s. From this point onwards, mastering changed massively. At that time, record labels owned studios, and the labels employed the engineers directly. Careers began at the bottom of the ladder as apprenticeships, and the first rung was to spend time with the mastering engineer. The idea was that apprentices would hone and develop critical-listening skills.

The mastering engineer's job was to transfer the final tapes from the mix/balance engineer, doing so as accurately as possible. The goal was to duplicate the sound of the tape on the disc. As an apprentice, engineers would listen to hundreds, if not thousands, of transfers. The huge benefit here was to spend time with an experienced professional. As experience and skills were gained, the apprentice stepped up the ladder to train with the mix engineer and then the recording engineer. Seemingly, the dark art of mastering that we alluded to yesterday is not so dark. The prized position was that of the recording engineer, a job that is hardly ever mentioned nowadays. Have you ever heard of someone talking about getting it right at the source? There we go.

As relationships between labels and studios fractured over time, engineers went freelance and began to work across multiple studios. This was challenging because each studio would have a unique mix environment. The job was always the same though: Polish the mix they had in front of them with the available tools (EQ, compression, and effects), albeit in a less familiar setting. This is effectively where we are today. The mastering engineer's role has developed to become the final quality control for not only technical aspects of the recording but also artistic ones too.

> TASK – Given the importance placed on the recording engineer, reflect upon whether there is more you can do to improve the quality of your records at the source.

Day 4 – Preparing to master

The art of getting a good master lies very much in preparing the audio file you're working on. If you prepare correctly to master, you'll set yourself up for success. So let me run you through eight points that you should always go through before mastering:

1. Optimise your listening environment. Trying to master in an untreated room is impossible. When you pay a professional to master for you, most of what you're paying for is the treatment of the room. If you only take one thing away from this chapter, make it this. The only way to guarantee that you are optimising your mix at the mastering stage is to do so in a room that will give you an honest reflection of your track. I won't go into room treatment here as that deserves a book of its own. But do some reading on the subject.[1]
2. Make sure your mix is finished. You can't master from an unfinished track. It's important to know that you're no longer worried about

individual components of the track but are now entirely focused on the sum of the parts.
3. Check your levels. I'll talk thoroughly about target levels tomorrow.
4. Bounce your mix down to a stereo file. You should bounce your mix out at the same sample rate and bit depth as the project file. So, if your session is at 48kHz 24-bit, then this is what you bounce at. You can master either .wav or .aiff files.
5. Take a break. Don't try to master your track the same day you finish mixing it. I'll talk more about this later too.
6. Import your stereo mix into a new project to master it. Don't be tempted to try to master it in the mix session. It's not good practice. The main reason for this is that you'll be tempted to go back and fiddle with mix elements, thereby taking your mind off the task you should be focused on: Mastering!
7. Listen through the song from start to finish and take notes. You'll identify most of its issues in this first listen.
8. Import your reference tracks into the session, and then make some A/B comparisons between them and your mix. Write down the main areas you need to address to make your master fit in with its target crowd.

Once you've done all this, you'll then be ready to master. As you can now see, a lot of work and effort goes into setting yourself up for success. I advise you to follow these steps diligently.

> TASK – Reflect on this eight-point list. How many of these have you been doing previously? Which points do you need to build into your workflow?

Day 5 – Mix levels and compression

At what level should you deliver your mix for mastering? This question gets a lot of airtime, and it can be challenging to find a definitive answer, perhaps because there isn't one.

What is vital is to leave some headroom. Headroom is space between your peak level and 0dBFS (full scale). Remember, 0dBFS is the point above which your audio will be clipped. You should ensure your mix peaks no higher than −3dBFS, with an RMS (average) level of −10/−14dB. Note there are two different levels to pay attention to here. You want to ensure that both your loudest peaks and average levels are in check. Your dynamic range may be too narrow if your peak level is at −3dBFS, but your RMS is at −8dB. This may well indicate an over-compressed mix.

Conversely, if you have a peak level of –3dBFS but an RMS of –18dB, your dynamic range may be too wide, perhaps indicating a lack of compression throughout your mix. Note that I say *may* speculatively. Everything is dependent upon the track and the context. The relationship between the peak level and RMS is unique to that track and should be considered as such. This relationship will look and sound very different for a jazz track compared with an EDM track.

Mix buss compression is the next piece of this puzzle. Should you include it in your bounce or not? If you've been mixing through a mix buss compressor, it may have become an integral part of your mix. The simple consideration is this: Anything that has been written into your stereo bounce cannot be undone. Therefore, best practice is to deliver two versions of your mix, one with the mix buss compression and one without. This will give your mastering engineer the option to use it or not, but at least it will provide them with information about the direction in which you are leaning.

A quick additional note at this point: Don't include fades in your mix. As with mix buss compression, these can't be undone. Leave them to your mastering engineer to do. Fades can be made shorter very easily but can't be made longer.

> TASK – Revisit some mixes you've finished. Analyse the relationship between peak and RMS levels. Where are your levels sitting? Is there a difference from track to track?

Day 6 – Loudness, the war, and penalties

The hottest topic of the 21st century regarding mastering levels is undoubtedly loudness. Everyone has an opinion on it. I've found this to be the subject with the most garbage said about it online. There are so many opinions, most of which are wrong.

The general misconception is that the louder your master is, the better. This opinion came about when people were trying to make their master stand out from the crowd on the radio. The reality is that tracks get compressed a lot before being broadcast. This is to protect consumer-grade listening devices. This broadcast compression clamps on a signal so much that songs that are already super loud get destroyed. You'll know this if you've ever heard a song you know well on the radio. Compared to your CD version, or even streamed version, it sounds very different. The louder the master is, the smaller it will sound on the radio.

A similar version of this still applies to the digital domain. Digital Streaming Platforms (DSPs) need a safeguarding method to ensure that one song doesn't appear drastically louder than another. Most listeners these days don't listen to an album from top to tail. They listen to a playlist containing songs from a range of artists. DSPs, therefore, employ loudness normalisation, meaning that all songs will be normalised to the same loudness level. This ensures a consistent listening experience for the consumer. Spotify's normalisation level is −14dB LUFS integrated, for example. If you deliver a master above this level, it will be turned down, making it quieter. Let's quantify this a little. You deliver your master to Spotify at −1dBTP, −8dB LUFS integrated. Therefore, Spotify turns your song down by 6 dB, which means your peak level becomes −7dB. This is called a loudness penalty. Therefore, a more dynamic mix at the same integrated LUFS level will sound punchier when streamed than an overly compressed or limited one.

The argument here has actually to do with perceived loudness, which is unmeasurable. How loud does your song *sound*, regardless of how loud it is on a meter? This is a much deeper rabbit hole, which I'll touch upon later.

The real takeaway from this is that, when mastering, you shouldn't be concerned with your loudness on a meter. You should master according to what works best for the song. If you distribute your tracks to a DSP, whether it's Spotify, Tidal, SoundCloud or any other, they'll all normalise your levels in one way or another. Likewise for any kind of radio broadcast. The only way to ensure your track sounds as intended is to sell a physical copy. And even then, the level at which the consumer listens to your record will affect how they perceive it. As we learned much earlier, low frequencies are much more perceivable at higher SPLs.

> TASK – Today's task is about shifting mindset. If you previously thought louder was better, get rid of that POV. Start thinking in terms of what best serves the song. What is the optimum loudness for *that* song?

Day 7 – The most important piece of gear is . . .

. . . the room. It doesn't matter how much analogue equipment you have or how expensive your monitors are. If you haven't got a good space to listen in, it's all for nothing. This is also important at the mixing stage but to a lesser extent. You can make a solid mix in a less-than-perfect listening environment, but the same can't be said of a master. Remember, mixing is like coaching the individual components and sections of a team. Mastering is like managing the whole squad. The room is like the experience good managers bring with them. The more experience managers have, the better their decisions are likely to be. A good mastering room tells you everything about your mix – both the good and the bad. A large part of this is to do with translatability. Part of being a mastering engineer is ensuring that the master translates across every listening medium, whether in a car, on earphones, a Bluetooth speaker, or anything else.

The size and proportions of a room play a significant part in how the room sounds. Various highly complex mathematical equations help calculate ratios between length, width, and height of walls and ceilings for better sound. If you want to explore the subject, investigate Oscar Bonello's research.[2]

Beyond the room's proportions is the treatment of the space. This comes in two forms: First, absorption, and secondly, diffusion. Absorption stops frequencies from reflecting to your listening position, meaning they don't interfere with the direct sound coming from your speakers. Diffusion works by scattering problematic reflections of sound in different directions. The theory behind acoustic treatment is complicated. If you want to treat your space correctly, I advise speaking to an acoustician for some proper advice. However, there are a lot of resources online that explain how you can do it reasonably well on a budget.

> TASK – Do some research regarding acoustic treatment. Learn where the important places are to place absorption and diffusion. Review your setup and consider if there are improvements you can make.

Day 8 – Don't master your own tracks

For the reasons explored yesterday, it often makes sense not to master your tracks but to pay someone else to do it. You're not just paying for the experience of a seasoned mastering engineer. You're also paying for their room. A good mastering engineer will work in a good space that is well treated. Be wary of any mastering engineer who isn't proud of their room. As mentioned previously, the room is *the most* crucial piece of kit.

You'll often be told not to master your own songs for a reason that has nothing to do with the room. It's a popular opinion that it can be challenging to make impartial decisions on a master when you've been involved in every stage of its creation to that point. Employing a mastering engineer allows a fresh pair of ears to bring the track together in an impartial, uninvested manner that is often beneficial to the track.

I don't sign up to this school of thought. I master all my material. But there are a couple of ground rules that I like to follow. Most importantly, always remember to do what's best for the track. This is part of my mantra from conception to distribution. Every decision I make throughout writing, recording, mixing, and mastering is made for the good of the song. Or at least this is what I aim for. This takes practice, of course. You must train yourself to depersonalise the experience, to approach it with a fresh pair of ears from a neutral stance.

Linked to this is taking some time between finishing your mix and mastering it. Don't master the day you hit print on the mix. There are a couple of reasons for this too. First, you'll want to check the mix on various playback systems. You're likely to want to make a couple of tweaks to it. There have been a number of times when I've had a client ask me to master a track, only to be told that they don't like the tone of the bass, or the vocal, or something else. Be 100% happy with your mix first. Secondly, you should give your ears some rest. You need to get the mix out of your ears for it to feel a little fresher when you come to master it. Leave it alone for a couple of days as a minimum, but a week or more if you can.

> TASK – Practise writing notes on what someone else's mix needs in the master. Get used to depersonalising the situation. Then apply this practice to your own mixes.

Day 9 – Digital vs analogue

Mastering can be done digitally, on analogue gear, or using a hybrid setup. One isn't better than the others. Many mastering engineers today use a digital setup because it's much easier to save and recall later. Working with analogue outboard gear is fun, but saving and recalling settings is a real pain, especially if the equipment doesn't have stepped pots.

The debate about analogue outboard gear sounding better, different, and warmer than its digital counterparts will go on forever. There will always be purists that think the real thing imparts a certain je ne sais quoi. And there will always be those more economically minded who believe their digital-replica counterparts do a damn good job. I fall somewhere in between. I love outboard gear and think it sounds different, but you won't catch me blowing 10k on a Fairchild. I have three emulations that do a fantastic job already.

Another benefit of working digitally is that this provides more flexible routing capabilities. It's easier to send anything anywhere you want when you keep things in the box.

If you're going to pay a mastering engineer, their setup will be factored into their cost. Those who work in the box should be cheaper than those who have outboard gear to maintain. Where you take your business is up to you.

If you wish to master yourself, I recommend you get some analogue emulations to work with. Great starting points are a Neve 1073 and an SSL EQ, a couple of buss-compressor options such as an SSL G or a Neve 33609, and a decent limiter such as FabFilter's Pro-L 2. Yes, there are far more complicated plugins to use for mastering, but these are great starting points that will give you excellent results without breaking the bank.

> TASK – Find some comparison videos of analogue vs digital gear and see if you can hear the difference. Review your plugin library and consider if there are things you need to supplement it with at this stage.

Day 10 – Optimising, *not* fixing

Let's get one thing straight here. There is no such thing as 'fixing it in the master'. Anyone that has ever said that doesn't understand what mastering is about. As we've already discussed, mastering is about pulling all the parts together effectively, making them work as effectively as possible as one cohesive unit. If you have an issue with any individual element in your mix, you won't be able magically to fix it in the master. The opposite is the case. Imagine you have a problem with your bass. Maybe it has too much low-mid content, making your mix muddy. If you wait until mastering to address this, when you try to remove the offending frequencies, your only option will be to remove that area from the whole song. This will leave your track sounding hollow. Or perhaps you realise you don't have enough high-end shine in your vocal. If you try to add this in your master, you'll have to add it across your whole track, making it overly bright and brittle.

Your mix should be perfectly listenable. Besides being quieter than commercially expected, you should be happy to listen to your mix from start to finish without hearing anything noticeably offensive. If this isn't the case, then your mix isn't finished. We mentioned earlier about waiting a little between finishing your mix and starting your master. This allows you to safeguard your mix and ensure it's ready for the mastering table.

When a mastering engineer (whether you or someone else) first listens to your mix, they shouldn't be interpreting it as something to make their own. Mastering isn't a creative process. It's more technical. The focus should be on optimising, not altering. Mastering is about making the most of what is there, not fundamentally changing the mix entirely.

The alterations made in mastering will be subtle and nuanced. There will be no tremendous EQ moves, large amounts of compression, or heavy saturation moves. Any significant movements such as these will fundamentally alter the overall character and feel of the mix, which could well negate the producer's or artist's intentions.

> TASK – Look back at previous mixes. Are there moments where you thought, 'I'll fix that in the master'? If so, fix them now. There's no such thing as fixing in the master!

Day 11 – Good mastering preparation

In addition to the list of points we covered a few days ago regarding preparing to master, there are several additional considerations if you choose to employ someone else to master for you.

First, ensure that your mix versions are labelled clearly. Make sure it's easy for the mastering engineer to identify which one is the unprocessed mix, which one has buss compression for reference, which one has limiting for reference, etc. The easiest way to annoy your mastering engineer is to give them files labelled 'bounce 1, 2, and 3'. It can be helpful to deliver multiple mix options, too – for example, a version with the vocal slightly higher in the mix and one slightly lower. Or one with the solo a little more cranked and one a little less. Again, ensure these options are all sensibly labelled.

This next part depends on what medium you're mastering for. If you're mastering for digital distribution, you can ignore this part. However, if you're mastering for a physical release on CD, you need to have titles, ISRC codes, metadata, and CD text at hand. All this information is written into the file during the mastering process when mastering for CD. However, if mastering for DSPs, all this information is usually added later as part of the submission through your distributor.

There are other further considerations to make if you are releasing physically, such as knowing your manufacturer's required format of delivery, ensuring your songs are accurately timed out to ensure they will physically fit on the release medium, supplying reference tones with pre-master mixes if on tape, and so on. But this is pretty niche territory these days and is well beyond the realms of most home producers.

> TASK – Reflect upon how you label files in general. Are there improvements you can make to how you organise things?

Day 12 – The four stages of mastering: EQ

Once you have done all the administrative preparation, you can get into actually mastering the music. There are only four stages here, the first of which is EQ.

As mentioned previously, the job of the master is not to make the mix sound overwhelmingly better. The mix should already sound great. The purpose is to help the mix sound its best on any listening device. It stands to reason that experience counts for a lot in this area. Experienced engineers will understand what areas to target to make the mix more transferable. This shouldn't deter you from doing it yourself. Everyone was inexperienced once.

Any EQ adjustment you make at this stage will be done very gently. Generally, you shouldn't boost or cut by more than 1dB at a time when mastering, perhaps 1.5dB at the most. This number is less than what others may tell you. I've heard 3dB used a lot as a maximum. For me, this is too much. If you have to adjust your master by that much to get it sitting where you want it to, you need to revisit your mix. And, of course, you should always use a reasonably broad bandwidth, so the move is subtle. Narrow subtractive cuts should be kept solely for mixing and should not enter your mind when mastering.

When applying EQ in a mastering context, you should use a linear EQ. The difference between a linear and non-linear EQ is challenging to articulate simply. Still, you can think of it like this: Linear EQs are designed to work on multiple instruments simultaneously. They're also clean, meaning they won't impart any additional colour as an analogue EQ would. That's not to say that you shouldn't use an analogue EQ on your master, but you should be careful with its implementation. You don't want to saturate the master to the point where you are recharacterising the song's overall feel.

Ensure that your EQ decisions are being made having regard to your reference tracks. Especially if mastering for digital release, the focus should be on ensuring the track will sit comfortably amongst other tracks in the same field. You want your song to feel as if it belongs amongst other pro tracks. You don't want it to stand out negatively for any reason.

Beyond these broad pieces of advice, some more specific matters can be mentioned. For example, you're safe to high pass up to 15Hz. Nothing this low is perceivable. You can high pass much higher in some genres, even up to somewhere around 50Hz in some cases. This will buy you headroom, allowing you to make your master louder in the long run.

You can use mid/side EQ to enhance the width of your mix. For example, a typical move is to high pass up to around 120Hz in the sides of your mix, focusing on the low end in the centre.

It's good practice to try to shape your mix subtractively rather than additively as much as possible. Consider it as making room for the frequency areas you want more of rather than taking away what you want less of. This adjustment in stance can be tricky to grasp but liberating when it clicks.

Finally, always gain stage. I've said it enough times by now, but don't be fooled by changes in volume. Louder will sound better. Match your input and output levels to ensure you're not being fooled.

> TASK – EQing a master is all about context. You should pay careful attention to your reference tracks to understand the context within which your track will sit. This will inform your EQ moves.

Day 13 – The four stages of mastering: Compression

The second stage of mastering is compression. Let's start by stating that this stage isn't a prerequisite. Whether your mix will need any additional compression will depend on the style of music you're working with, and how compressed the mix is in the first place. So don't assume that compression is always going to be needed.

Compression in mastering is used for two purposes. First, to control dynamics. Secondly, to add colour. Let's cover the latter first. As mentioned previously, adding colour is another way of saying adding saturation or distortion. These terms all mean the same thing. By adding colour to your mix, you'll enrich the harmonic content within, softening the transient content and, therefore, by association, glueing the mix together a little. The further you push the saturation, the more glued it will become, but also the more distorted. The line is fine here.

Some compressors are renowned for the colour they impart on a mix. The SSL-G is one. The Fairchild Variable Mu is another. It is common simply to pass the signal through these analogue circuits without applying any gain reduction merely to benefit from the colour the circuitry imparts. You can also benefit from this effect in the digital domain with good analogue emulations.

Compression's first use, gain reduction, is less easy to reduce into such a simple chunk of information. This is because there are so many ways to compress a master. However, one thing is agreed upon by almost all: Your gain reduction amount should be relatively small. 1–2dB is plenty, with 3dB being roughly the limit of what is perceived as acceptable. This covers your threshold. However, the other parameters on a compressor (ratio, attack, release, and knee) are less generally acknowledged. And I'm sorry to say I haven't got the golden egg that will provide you with the answer.

Again, this is because there isn't one size that fits all. But let's work through them logically. The ratio isn't too tough to work out. Mastering is about subtly shaping the sound, not smashing it over the head. So, a gentler ratio will be appropriate. Anything from around 1.5:1 up to a max of 4:1 could work. How high the ratio is will again depend upon the track.

Attack and release controls will arguably shape your sound the most. The slower the attack time, the more transient content will be allowed through before the compression kicks in. Some will advise you to start with a slow attack time and gradually roll it back until you're losing the punch you desire, then back off a little. This can work. However, I frequently set attack times almost as fast as they'll go to grab pokey snare transients as much as possible. Part of your consideration here also needs to be the compressor you're employing. Some respond much faster than others, so you need to know your tools and employ the right one for the job.

Release times are similarly opaque. You'll frequently be advised to start with a fast release time and then back it off until the release just returns to zero by the time the next percussive hit strikes and then back off a little. This method aims to get the compressor breathing in time with the music. Again, I often use the fastest release possible, so I'm just working on pulling the transients back into the mix.

Finally, the knee. Not all compressors have a knee, so this may be redundant information. But my advice here would be to start with a hard knee, dial in the rest of the settings, and then roll back the hardness slowly until you feel it's become too soft, and then back off again. Remember, the knee denotes how the compressor responds to the signal that exceeds the threshold.

Having said all this, I suggest you keep in mind the concept we've mentioned many times previously about having intent. Don't just stick a compressor on your master and fiddle until you think it sounds good. Listen critically to the master and form an opinion as to what, if anything, it needs. Does it lack punch? Is its dynamic range too wide? Know what it is you're setting out to achieve first. By doing this, you're more likely to end up somewhere that you like.

> TASK – Just as with EQ, mastering compression is all about context. How compressed are your reference tracks? Are they aggressive? Fast/Slow?

Day 14 – The four stages of mastering: Limiting

Limiting is the third stage in your mastering process. Again, this isn't a prerequisite either, although you'll certainly find it on all modern productions.

Limiting is the stage where you bring the loudness of your track up to a more competitive level. But don't be fooled. The uninformed believe that limiting harder and harder will make your track louder. This is not the case. Loudness comes from a well-crafted mix. You cannot achieve high perceived loudness levels just by slamming your mix with a limiter. If it were that easy, everyone would do it!

Here is something I don't understand: Why do some people tell you with such confidence to set your limiter's output to −0.1dB? or −0.5dB? DSPs are consistent in stating that they want tracks submitted at −1dBTP. Spotify,

Apple Music, Deezer, YouTube, and SoundCloud all say the same. So, if you're mastering for digital release, −1dBTP should be your output ceiling. The true peak part is essential. Ensure your limiter is set to monitor true peak levels.

The process of limiting is the same as compressing: You have a threshold, attack, and release control. However, the ratio is fixed at ∞:1. This means that any signal exceeding your threshold will be chopped off completely. This is what prevents clipping. For this reason, you should be extra cautious. Again, I recommend not exceeding 3dB of gain reduction here. The more controlled your dynamics are before they reach the limiter, the lower the threshold will be able to go before limiting occurs, and thus the louder your mix will become. Controlling your dynamics for loudness should happen at every stage of your mix, not just at your final limiter. You should leave your final limiter as little to do as possible to keep your master sounding natural and not obviously slammed by a limiter.

As with compression, the same question marks about the attack and release time apply. These will be genre dependent. However, bear in mind that they only really matter if they are responding to transient information. If your mix is slammed so hard that all the signal exceeds the threshold, your attack and release times become meaningless.

The one thing worth mentioning with release times is that times shorter than around 30ms will tend to introduce distortion to your mix. So, in general, try to keep your release time above this point.

> TASK – How loud are your reference tracks? Can you limit your track up to a comparable level without destroying its dynamics? If not, you should revisit your mix.

Day 15 – The four stages of mastering: Metering

Lastly, and this stage is *not* optional, is metering. To ensure your track will be received well on its intended medium, you need to meter.

Let's start with the obvious things. We've already mentioned loudness a lot, but let's go further. DSPs ask for a loudness level of around −13 to −15dB LUFS integrated. That number is measured by playing your whole song through the meter from start to finish so that the entire track can be measured. If you're releasing an EP or an album, the measurement will be taken across all tracks within the release. If you're mastering for CD, your target should be around −9dB LUFS int. And for club play, you should shoot for about −6 to −9dB LUFS int.

True peak is the next thing to meter. For DSPs, −1dBTP is your target. For CD and club play, you can shoot for −0.1dBTP. Both loudness and true peak values are talked about a fair bit.

What's hardly ever talked about is dynamic range (DR). The dynamic range of your song is the difference between the loudest and the quietest sections. A low DR value denotes a track that is over-compressed or over-limited. Generally, aiming for a minimum DR of 9dB is good. You can push this to 8dB for club mastering. Ensure you have some dynamic difference between the different sections of your song, and you'll be heading in the right direction.

Swinging back around to loudness for a moment, it's essential to have your different LUFS values separated in your mind. You can hit a short-term LUFS level of −8 or −9dB in your chorus or drop and still hit the −14dB LUFS int. so long as you have enough dynamic range in your song.

When mastering, be sure to keep an eye on your correlation meter. Although you've been paying it attention all the way through your mixing process, there's never been a more critical time than now. The last thing you want on a master is phase cancellation!

A decent metering plugin will monitor all these things for you. There are a few on the market.

> TASK – If you haven't already got a good metering plugin, get one! Make sure you know where to find all these critical pieces of information: LUFS, dBTP, DR, and correlation.

Day 16 – Using appropriate reference tracks

I've talked a lot already about the importance of referencing. But we haven't discussed how you go about selecting appropriate reference material. Let's break it down into six steps:

1. The first thing to do is to make a clear difference in your head between a good reference track and just something you happen to like. Just because you like the hook or melody, or think it has a great groove, doesn't mean that it's well produced. Being a good song doesn't automatically make it a good reference track. You should be looking for something that sounds great across all listening devices. This is a good benchmark test.
2. Use current tracks. Comparing a track produced in 2022 with something from 1982 is of little value. Unless, that is, you're aiming to recreate a specific vintage vibe. If you want to sound current, you must use current

reference material. This is even more important in fast-moving genres such as techno and house, where the landscape can move every few months.
3. Select genre-appropriate tracks. If you're aiming to make a club track, don't reference a pop-rock track made for radio. Every genre and subgenre has its characteristics and conventions. Referring to something unrelated will only get you in trouble.
4. You can get into hot water if you use signature tracks as your reference material. Some sounds are so synonymous with a specific artist that you can end up sounding like a copycat if you reference them too closely. You don't want to be accused of ripping off someone else. When referencing, it's better to select well-produced but more generic tracks.
5. Avoid busy reference tracks. The more clutter there is in a song, the less easy it is for you to identify what is going on. By referencing something a little sparser, you will more easily be able to differentiate between elements.
6. Use references with dynamic variety. There's very little to be learnt from something that is destroyed with a limiter. By using something with its dynamics intact, you'll learn more about the construction along the way.

TASK – Review the reference tracks you've been selecting up to now. Are they appropriate, or could you be choosing more useful references?

Day 17 – Trimming the start and end

Here's another topic upon which nobody wants to give any information: Trimming the start and end of your master. It's not a complicated thing, but it needs to be done correctly for your track to sound professional.

Automating your fades should be done on your master fader. This will ensure that it happens post any plugins you have, ensuring that no sound can creep through. You're trying to achieve complete silence at the top and tail of your song here.

The front end of the song is easy. You should zoom in close to the waveform and then automate your master fader just before the audio begins with a quick fade. Ensure your fade happens just before the track's audio starts so that you don't cut off the front end of the sound. This fade should be very short. You want the song to play almost immediately when someone hits play. You don't want an awkward pause at the beginning of your track.

The end is slightly more subjective. My method is this: Listen for the point at which the song has just finished, or the resonant tails of anything ringing out, crashes, chokes, etc., are just decaying. At this point, automate a fade lasting roughly five seconds. This should give your song enough room to breathe at the end before the next song plays. The exception here is that if you're mastering songs that need to flow into one another as part of an EP or album, you will need to make the adjustments at this point. So, if it's meant to run almost immediately into the next, your ending fade will be shorter and tighter.

Those who are a little extra like putting curves on their beginning and ending fades. I'm not one of them, but you may be! In which case, your intro fade should be logarithmic, and your outro fade should be exponential.

> TASK – Practise putting fades on some tracks. Ensure you don't cut off any audio. Make the ending feel natural. Add curves if you wish.

Day 18 – Bouncing/rendering/exporting

Bouncing, rendering, and exporting all mean the same. It depends on what DAW you work in as to what it's called. To do it is simple. Set a cycle that goes just beyond the fade points you created, so you're bouncing from and to 100% silence.

That's the simple bit. Where people get in a mess is what file formats to export with. Let's start with the simple stuff: Don't bother with .mp3! .mp3 is a lossy file format, meaning that the quality of the data will be degraded as the information is compressed to make the actual file small. What is the point in going to all that effort to make the best-sounding track you can just so that you can throw it all down the toilet at the last moment when you bounce it? You should be exporting as .wav. This is a lossless format, meaning the quality of the information kept in the file isn't degraded as the file is compressed.

The next consideration is to do with sample rate and bit-depth. If you've recorded at and kept your project at 48kHz 24-bit, then bounce at that too. More and more platforms now support HD audio, and those that don't will catch up soon. Tidal was the first to make a big deal of offering this. Apple Music is now on board, and others are sure to follow (as of spring 2023). Even SoundCloud supports it if you have a Pro package.

If your project is at 44.1kHz 24-bit, then that's fine too. Export at that. However, the general standard for CD is 44.1kHz 16-bit. If you are reducing your bit-depth, you'll need to include what's called dithering in your bounce. Dithering is low-level noise that's added to your files when bouncing down to a lower bit-depth to mask the audio quantisation. I won't baffle you with the reasoning but be sure that it's necessary. Note that you only need to dither if you're reducing your bit-depth. It's not required in any other circumstance. Also, note that there's nothing to be gained from increasing the sample rate at this stage or any other point. You cannot increase the fidelity of something. So don't bother bouncing your 44.1kHz session at 48kHz. It's completely meaningless![3]

Another important note: If mastering for CD, bounce as .aiff format. The explanation is in tomorrow's lesson.

> TASK – Review how you've been bouncing your tracks. Have you been formatting correctly?

Day 19 – Metadata

Metadata is information that is added to a music file to identify and present the audio. This information is vital as, without it, the music wouldn't be correctly attributed to the relevant parties, and therefore the correct royalties wouldn't be paid.

There are a whole host of details required as metadata. These include artist name, track title, album/release title, genre, songwriter credits, and track numbers. This information is straightforward, so you should have it on hand. You'll also credit any additional artists, producers, writers, or engineers involved at this stage.

One of the most critical pieces of data is the International Standards Recording Code, or ISRC. This code is your track's digital fingerprint and is used to track your song's plays and sales through DSPs. Your song cannot be distributed digitally without it. Don't fret, though. Most online distribution services will assign ISRCs for you if you're self-releasing. Note that the ISRC is for the song, not the release. So, every song will have its own code.

If you're only releasing digitally, then you don't need to do anything about adding metadata to the bounce in your DAW. All the metadata handling will happen when submitting through your distributor. However, if you're mastering for CD, you'll need to include it. Metadata can be written into .mp3, .flac, and .aiff file formats, but cannot be written into a .wav file, so bounce as .aiff in this situation. You'll also need to create what's called a Disc Description Protocol or DDP file. The DDP file is a precise electronic version of your music that is immediately ready for duplication. Not all DAWs support DDP file creation, but you can create it in both Reaper and Studio One.

> TASK – Start a spreadsheet. On it, store all of the metadata for all of your music. Add to it as you write and keep it up to date. You'll thank me in ten years when you need to refer back to something old.

Day 20 – Beyond the essentials: Saturation

At this point, you can master a track perfectly well. You can satisfy all the culturally expected and technically required aspects of mastering. However, there are a few additional tools that you can add to your chain that will take your masters a step up.

The first of these extras is saturation. We've mentioned saturation a lot already. We know that saturation is just distortion. Distortion, in moderation, is a good thing. Remember, when applied tastefully, it will gently round over transient peaks, subtly reducing the dynamic range, and will enrich the harmonic content, making your track sound thicker and louder. All these characteristics are good things when it comes to mastering.

How you add this saturation is the key here. You have several different options that I'll run through now.

1. The most common in the analogue domain is tape saturation. Simply by recording your master onto tape, you will benefit from the saturation that the tape imparts. In the digital realm, you can place a tape emulation plugin into your mastering chain before your limiter. There are loads of options out there that emulate all sorts of different models. You'll need to do some reading to figure out what model will be most appropriate for your sound. Don't drive it too hard. Subtlety is the objective here.
2. Tube saturation is next. It is renowned for sounding full and warm as it adds strong second-order harmonics. Creating an all-tube signal chain would cost a fortune in the analogue world. Digitally, you can get plugins that just impart the tube vibe, or you can get it as part of a saturation plugin like FabFilter's Saturn 2.
3. Other analogue emulations will have different circuitries built into them. For example, the Shadow Hills Mastering Compressor, which is used by many, has three different circuit options, each supplying its own flavour of distortion.
4. Analogue summing is an option you can do in or out of the box. I love my Neve 8816. It's a cost-efficient way of getting an analogue vibe in a hybrid scenario that doesn't break the bank. There are plenty of excellent hardware summing options out there. But you can also get analogue summing options that are applied within the box. Waves' NLS is an excellent option, for example.

When saturating your master, remember, as with anything, that you can automate it. Don't just set and forget. Adjust the amount of saturation to add additional weight to your choruses.

When adding saturation in whatever form you choose, look for an oversampling option and set it as high as it will go. I won't bore you with the technicalities here, but it's important. Also, try to avoid saturating high frequencies. Saturating high frequencies will sound harsh and aggressive and make them too loud. It can also make your transients too pokey, causing ear fatigue more quickly.[4]

> TASK – Explore some different methods of saturation on your masters. Start with some inexpensive options like virtual summing or virtual tape.

Day 21 – Beyond the essentials: Stereo enhancement

The next thing that is common to play around with when mastering is stereo enhancement. Don't be fooled. This doesn't just mean making your mix sound wider. That's certainly part of it, but making some parts of a song narrower will also make the other parts sound wider by association. Let's look at the subject in detail.

The objective of stereo enhancement is to make a mix sound wide and open, making each element in the mix clearer and more perceivable. There are lots of plugins made exclusively for this purpose. iZotope's Stereo Imager is probably the most famous in the digital domain. But it's also found on hardware units such as the Neve 8816 Summing Mixer, which is renowned for adding width to any mix that passes through it.

If your mix sounds a bit dense and lacks definition between parts, stereo enhancement is an excellent place to look for a solution. However, you should always proceed with caution. Being too heavy-handed with stereo enhancement is likely to introduce phasing issues. This means that when the mix is played in mono, it'll sound worse than if you'd not fiddled with your stereo image at all.

This is why I mentioned narrowing as well as widening a mix at this stage. Narrowing a mix won't add any phase issues, so it's less likely to have a damaging effect than widening. It may shine a light on any phase issues already present in your mix, but it won't create any new ones.

Let's be clear: We're not talking about narrowing or widening by 20 or even 10%. We're talking small numbers like 5% that are almost imperceivable. You don't want to notice a mix getting wider at any point.

Another good tool that doesn't risk adding or enhancing preexisting phase issues is mid/side EQ. By removing some low-frequency content from the sides of your stereo image and enhancing the high end, you can create the feeling of widening your mix. Aim to high-pass around 100–120Hz and add a high shelf around 7–8,000kHz. You can also remove some top end from the mid channel and even subtly boost the lows to complement the moves you make on the sides.

> TASK – Investigate some stereo enhancement options. Try automating a stereo imaging plugin, as well as some mid/side EQ.

Day 22 – Beyond the essentials: Parallel processing

Next on the list of optional mastering processes is parallel processing.

The first thing that jumps to mind when considering parallel processing is compression. This is primarily considered in the mastering world as a way to increase the overall loudness of a track. By sending your entire mix to a buss and hitting it hard with a compressor, you create a dynamically flat version of your mix. Blending it back in underneath your main mix will increase the overall RMS level, leaving you with a louder mix without impacting your peak levels. When done tastefully, this can be great. When done poorly, it will destroy the dynamic range of your mix, leaving you with something flat and uninteresting.

The second option that is worth playing with is parallel saturation. You may have already experimented with adding some saturation directly into your mastering chain. If you found that this felt a bit heavy-handed, then applying it in parallel could be another route to take. You still have all saturation options available, so you can use your various shades of tube, tape, and amp distortion to thicken your mix. The advantage of applying it in parallel is that you can be a little more heavy-handed in your application, knowing that you will blend it underneath your main track. Beware of the same pitfalls here. Having your parallel saturation too high in the mix will destroy your dynamic range.

Another fun option is to experiment with some parallel chorus. Chorus has the effect of thickening through its subtle detuning qualities. Something like the Roland Dimension D SSD-320 is excellent for this. There are some great plugin emulations available. Overloud's version is fantastic.

With any parallel processing you do on your master, don't just set and forget. Automate your parallel levels to come in and out to enhance sections of your track. Increasing the loudness of your chorus or drop by pushing the parallel compression or saturation can work well, but it will be less effective if you set a level and leave it at that throughout.

> TASK – Experiment with some parallel-processing options. There are lots of options here. Try to discern their differences and identify what is better suited to certain situations than others.

Day 23 – Beyond the essentials: Serial limiting

Serial limiting is a good trick to know if you need to make your master loud. So, it's a technique that is particularly applicable to club masters and electronic dance music in general.

The need to make your master loud to compete with other tracks may tempt you to hit your limiter hard. The problem is that the harder you hit your limiter, the more likely you will encounter distortion. Any distortion caused by limiting a master isn't a good thing, so it should be avoided at all costs. Obviously, the preferable route is to find other ways to make your master loud, as limiting, by definition, will take out more dynamic range the harder you hit it, thus making your track less punchy, less dynamic, and therefore less engaging.

As I said previously, my magic number for limiting is −3dB. I try not to exceed 3dB of gain reduction on a limiter. But if this just isn't enough, you can employ two limiters in series. By limiting up to −3dB of gain reduction twice rather than trying to do −6dB in one go, you're more likely to avoid distortion. In this way, you could potentially explore slightly different attack and release times, so the limiters respond to the transient content differently. That's not to say that you can't get away with limiting hard in one go. You may get lucky. But gently narrowing the dynamic range of something in series rather than slamming it in one shot will always sound more natural and less perceivable.

The critical thing to remember here is to have your first limiter's ceiling set at 0dB and the second at −1dBTB. This will save you an extra dB of headroom. Every little helps when you're trying to achieve maximum loudness!

> TASK – Try serial limiting. Compare it to more aggressive limiting on a single plugin. Can you hear the difference?

Day 24 – Adding reverb

It may surprise you to know that you can also add reverb to a master. Let me immediately quantify that by saying this is by no means a necessity, but it can make all the difference when done sparingly and tastefully.

There are two main goals you may set out to achieve when adding reverb to a master. First, to add depth and contrast. Reverb is used to create depth, which implies distance. Using reverb for depth on a master is a great way to establish further contrast between sections. You may add more reverb to intro sections, verses, or bridges to make them slightly more distant, dropping the level for the chorus so it feels more up-front and in your face.

The second goal may be to add width and ambience. Reverb creates space. Therefore, it can be used to glue parts of a master together by locating them similarly. Large reverbs like halls will also provide additional width to your track. Again, automating the reverb level is key to delivering contrast between sections.

Your master's reverb should be placed on a buss, like a parallel process. This way, you can easily automate the buss's level as desired throughout the track. As with any effect on a buss, ensure the reverb is 100% wet. Ensure that you EQ your reverb here too. You can easily wash out your low end if you leave too much in, or add too much shine making your track feel brittle. Use a high-pass filter somewhere around 500Hz and a low-pass around 10kHz. You may also want to keep your presence range tidy with a cut somewhere between 2–5kHz.

The issues to look out for when adding reverb to a master are frequency masking and/or build-ups, comb filtering, and loss of clarity. The best way to avoid all of these is to be subtle. Remember, if you can obviously notice the reverb in the mix, it's probably too loud.

> TASK – Try adding reverb to a master. Experiment with different reverb types. Automate your send amount between different song sections.

Day 25 – Avoid ear fatigue

I know I'm going to sound like a broken record for repeating this but avoid ear fatigue! You want your ears to be fresh as much as possible when mastering. We mentioned earlier about ensuring that you have a period between finishing your mix and beginning your master. It's also important not to work on the same master for too long. Unlike mixing, where there are many elements to draw your attention at any one time, when mastering, you're focusing on the whole track. Therefore, you're only actually listening to one thing. If you find yourself working on a master for more than one hour, I recommend walking away and coming back the following day, or at least a few hours later. You want to spend as much time as possible in front of your master from the position of a brand-new listener, someone who's never heard the song before. When you become too familiar with a track, you overlook the detail. By keeping your ears fresh, you maintain the ability to focus on the small print for longer.

An important thing to mention here is that you don't need to listen loudly throughout the entire mastering process. When applying your EQ, compression and limiting, you'll certainly want to listen at a reasonable level, particularly when referencing other tracks. You'll want to 'feel' the music. But for all the technical aspects of mastering, of which there are plenty, you don't need to listen loudly. When you're metering, working out how to get your track sitting at your target LUFS level, automating, or anything else technical, turn the volume down. Give your ears some rest. This will allow you to maintain focus for longer and prevent you from going ear-blind.

> TASK – Practise turning your track down for periods of time. Think about what level you need to listen at, and turn it down when you can.

Day 26 – Stem mastering

There is a scenario that sits somewhere between mixing and mastering, and it's called stem mastering. It effectively creates an additional stepping stone on the journey to a release-ready track.

A stem refers to a logical group of instruments that have been exported together into one stereo file to reduce track count. For example, you may bounce all your drums down into one drum stem, all your backing vocals to another, all your guitars to another, and so on. So, rather than being left with one stereo file once you've finished your mix, you're left with however many stems you opted to reduce your track to.

Stem mastering has its pros and cons. While developing your skills as a mixing and mastering engineer, it gives you an extra safety net. You retain a little flexibility during the mastering stage, so if you find anything you need to try and address in your mix, you can hopefully fix it from the stem rather than going all the way back to the mix session. Undoubtedly, this is a good thing whilst you are honing your skills. The ability to redress a specific mix element from your mastering session can alleviate some of the stress of committing to a single stereo file.

On the other hand, it can reduce your focus whilst mixing. If in the back of your mind you know that you don't need to be fully satisfied with your mix because you can still work on it in the master, you may find yourself less attentive. This isn't what you want. The other major negative, as I see it, is the blurring of lines between the different processes. I believe it's important to commit to a mix before moving on to mastering it. If you don't fully commit to a mix, you will never know when it's actually done. You may effectively treat your master as a second mix rather than a master in its own right. I think keeping all stages in the music-production process separate is important. Writing is independent of producing, recording, mixing, and mastering.

Some mastering engineers prefer to work from stems. Perhaps this is because they've been sent so many poor mixes to master over the years that they've learned that their life is made easier if they have the stems to work from so they can correct any howling mix errors themselves. My POV differs from this. I believe that good mastering is about a conversation with the client. As a mastering engineer, you should understand what the end goal is for the track. And you retain the ability to communicate any mix issues with the client.

> TASK – Give stem mastering a go. Do you prefer it to regular mastering? Does it feel unnecessary? Make up your own mind!

Day 27 – Considering the medium and technical limitations

Whilst we've discussed the technical limitations of mastering for DSPs, CDs, and the club, I've made one major oversight here. Vinyl has made a significant comeback of late, so it would be remiss of me not to inform you of some of its limitations. Whilst 99% of your mastering will most likely be for digital release, you may be lucky enough to cut something to vinyl one day. If you're not well informed, this process will be a massive letdown!

In its simplest form, vinyl is a groove cut into a record read by a stylus. That stylus can measure variations in the groove down to one micron (1/1,000th of a millimetre), so it's a highly sensitive system! If the stylus doesn't like how the groove is cut, it could skip or distort. So, you need to know what you're doing.

Now, I'm not going to summarise how to master effectively for vinyl here. There's just too much to it. And it would offend mastering engineers everywhere if I attempt to do so! However, it is perhaps by understanding some of the significant challenges that mastering for vinyl poses that we can begin to appreciate why the mastering engineer was held in such high regard for so long. I'll cover some of the headlines briefly.

The loudness of vinyl depends on the length of the material that's cut onto it. The more music there is on a record, the closer the grooves need to be cut together to fit it all on. In this case, the level needs to be turned down and the bass reduced to get everything to physically fit on.

Vinyl spins at a consistent speed. A 12" LP revolves at 33.3 RPM. Thinking logically, this means the inside grooves move faster than the outside ones. Almost twice as fast, in fact. So, they have to fit the same amount of musical content in half the space. How does this affect the sound? By losing more and more high-frequency content the closer to the centre of the record you get.

For this reason, it's recommended to put faster, more energetic songs on the outside of a record and slower tracks and ballads towards the inside.

You need to balance the lengths of your sides. If you have one side shorter than the other, it will still be subject to the same accommodations that have been made for the longer side. Your short side will also be louder.

Avoid wide stereo bass when mastering for vinyl. When cut into a record, it can cause the needle to jump out of the groove. The problem is compounded if there is phase cancellation between the left and right channels of the bass. Sibilance also causes additional issues on vinyl. Excessive sibilance will cause distortion. The excess of high-frequency content at a relatively high level will result in the stylus being unable to track the groove accurately. Thus, distortion.

Through the decline in physical and the increase in digital releases, the need to understand the intricacies of the medium almost disappeared. The technical requirements of DSPs are far more consistent and predictable than the relationship between record and stylus. This is precisely why you shouldn't fear mastering your own music. Assuming you're mastering for digital release, CD, or club, you can absolutely do it yourself. However, if you're considering cutting a vinyl, maybe consider engaging a professional. Expertise in this area really can't be learned in a book!

> TASK – There's no real task today. Just a general appreciation for the subtle nuances involved in mastering for vinyl.

Day 28 – Unit summary

That's it! You've reached the end of the unit and the end of this course. The most daunting subjects, in my opinion, are those of synthesis and mastering, so if you're reading this, congratulations. You gritted your teeth, put in the hard yards, and have made it here.

Let's do a quick recap of what we've just covered. We've discussed:

- The differences between mixing and mastering
- A brief history of mastering
- Preparing to master
- Mix levels and master compression
- Loudness
- Acoustic treatment
- Digital vs analogue
- Optimising, not fixing
- The four stages of mastering: EQ, compression, limiting, and metering

- Additional processes such as parallel compression, saturation, and effects
- Avoiding ear fatigue
- Stem mastering
- Mastering for vinyl

I'm sorry to break it to you, but now comes the hard part! I hope over the past 12 months you've been applying your newfound knowledge, day by day, bit by bit. Now you need to consolidate that knowledge. That probably means revisiting various chapters that feel less familiar and doing some additional reading in areas that particularly spark your interest.

This book is a thorough introduction to what I consider to be the key aspects of producing music the right way. But there is a lot more detail to be unearthed. I hope your enthusiasm has been sparked and that you're now making music consistently on a daily basis. Good luck, and happy music-making!

Further reading

1 Mayes-Wright, C. (2009). *A beginner's guide to acoustic treatment.* [online] soundonsound.com. Available at www.soundonsound.com/sound-advice/beginners-guide-acoustic-treatment [Accessed 9 Nov. 2022].
2 Foley, D. (2015). *Ideal room size ratios and how to apply the Bonello graph.* [online] acousticfields.com. Available at www.acousticfields.com/ideal-room-size-ratios-apply-bonello-graph/ [Accessed 9 Nov. 2022].
3 Keeley, E. (2021). *What is dithering? The ultimate guide for beginners.* [online] emastered.com. Available at https://emastered.com/blog/what-is-dithering-audio [Accessed 9 Nov. 2022].
4 Mantione, P. (2021). *Oversampling in digital audio: What is it and when should you use it?* [online] theproaudiofiles.com. Available at https://theproaudiofiles.com/oversampling/ [Accessed 9 Nov. 2022].

Index

1176 69, 74, 165–166, 168

acoustic treatment 199, 203; and absorption 203; diffusion 203
additive/creative EQ 49–50, 52–53, 57, 164–165
ADSR 128, 182
ambience collision 102–103
analogue EQ 58, 61–62
API 53, 62, 68, 165
arpeggiator 185
arrangement 193
attack 66–67, 75–76
automation: binary, step, and spike 129; delay 121; fades and curves 128–129; panning 35–36; vocals 171
automation modes: latch 128; read 127–128; touch 128; write 128
auto-pan 146
aux send 93, 95, 99, 138–139

backing vocal 169–170
balance meter 34, 36
bit depth 3–4, 6, 213
Bonham, John 89
bucket brigade 109
buss 18; and drum buss 134; master buss 147–148

choke 142
chord trigger 185
chorus 114–115, 120, 217
click track 148–149
clockface technique 11–12, 30
Collins, Phil 91
comb filter 98, 108, 115–116
comping 152, 161

complementary panning 33–34
compression 165–167
correlation meter 13, 20, 36, 211
crosstalk 37–38
cutoff 141

de-esser 137, 167–168
delay 107–108, 139; and analogue 109; digital 109–110; doubling echo 111; dub 113; Grand Canyon echo 108; looping 111; multitap 112; ping-pong 112–113; slapback 108, 110; tape 108–109
digital signal processing (DSP) 156
direction mixer 122, 135
direct sound 85–86
Disk Description Protocol (DDP) 214
distortion 81, 139, 168–169, 208
dithering 213
double-track 29, 31, 33, 35, 37, 101
downshifter 144
dynamic EQ 45–47, 61, 102, 137
dynamic range 64, 211

ear fatigue 4, 219–220
echo 98, 107
envelope 141, 181–183
equaliser: shelf 48; bandwidth/Q 48; bell 48
exponential 129, 213
extreme-threshold technique 75–76

Fairchild 70, 166
FET compression 69, 76–78
FFT meter 20
filter 181; and all-pass 116; high-pass 48; low-pass 48; slope 48

flamming delays 122
flanger 114–115
flutter 109; and flutter echo 89; tape flutter 113
freeze 131
frequency 43–45
frequency allocation 55, 79–80, 172
frequency spectrum 51–52
full scale peak meter 6–7, 19
fundamental 43–44, 81

gain staging 5–8, 49–50, 55–56, 67, 73, 78
Glyn Johns method 26, 38
graphic EQ 45–47

Haas effect 94, 108
Hammond organ 90, 189
hardcore 164
harmonic 81
Harmonic Overtone Series 44
headphones 153–154; closed-back 153–154; and open-back 153–154
headroom 200–201
high-pass 102; and high-pass filter 25, 28, 32, 53–54, 58–60, 80
hip-hop 164

impact 144
impulse response 91
indirect sound 85–86
insert 72, 93, 95, 99
interleaved 26
isolation 153; and isolation booth 153
ISRC 214

jazz 164

Karplus-Strong synthesis 190–191
knee 67

LA-2A/3A 69, 75, 166
latency 156
LCR technique 31–32
LFO 114–116, 146, 184
linear phase EQ 58, 207
logarithmic 129, 213
loudness 201–202, 210; and loudness meter 19–20; loudness penalty 202
low-pass 102; and low-pass filter 28, 32, 54, 109
LUFS 19, 210–211

make-up gain 67
Manley 70
mastering 197–200; and compression 208–209; EQ 206–207; limiting 209–210, 218; metering 210–211; saturation 215–216
metadata 214
metal 164
microphone emulation 155
MIDI 170–171
mid/side 13; and mid/side EQ 25, 28, 207, 216
mixing hierarchy 15, 17
mix translation 60
modulation 113–116, 183; and modulation matrix 185
mono compatibility 13, 20, 122; and mono EQ 60
MSEG 185
multiband compression 75, 80, 82

Neve 53, 62, 68, 165, 204–205, 215–216
noise floor 3
noise gate 91–92, 160–161
noise generator 185
non-linear 68–69
normalising region gain 7
notch 115–116

optical compression 68–69, 76–77
orchestration 193
oscillator 141, 176–179; and sub oscillator 185
overdrive 168
overtone 43–44, 178

pan: bass 28; delays 122–123; drum panning 25–27, 34; guitars 28–29; horns, wind, and strings 30; orchestral panning 39–40; pan pots 23–24; synths and keys 29–30; vintage vibe 38–39; vocals 31
parallel compression 72, 139, 142, 167, 217
parametric EQ 45–47, 61
perceived loudness 3
perspective 24, 30
phase 13, 26, 98, 114; and phase cancellation 9, 13, 37
phaser 114, 116
piano roll 146
pink noise 16–17, 185
plosives 137, 154

polarity 9, 16
poles 116
pop 164
pop shield 154–155
post-fader 101; and post-fader metering 8
pre-fader 101, 158; and pre-fader metering 8
proximity effect 156–157, 163
Pultec EQP-1A 53, 58, 61–62

ratio 65
reference track 4–5, 211–212
release 66–67, 75–76
resonance 115, 141, 181
resynthesis 189
reverb 139, 218–219; and ambience 92; cathedral 92; chamber 88–89, 91; convolution 91; gated 91–92; hall 88, 92, 103–104; non-linear 93; plate 89–90, 94; reverse 92–93; room 89, 92, 103–104; shimmer 92; spring 90, 94
reverb parameters: decay 93, 96–97, 103–104; diffusion 94, 98–99; early reflections 94, 98; pre-delay 94, 97, 103–104; size 93, 96–97
reverb tail 90–91
riser 92, 140, 144
RMS 200, 217
RnB 164
rock 164
Roland Space Echo 113
ROMpler 188

Sabine's Equation 87
sample 25, 27, 109, 146; and sampler 187–188
sample rate 213
saturation 81, 139, 168–169, 208, 215–216; and parallel 217; summing 215; tape 215; tube 215
sawtooth wave 178–179
serial compression 74–75, 166
series 72
shelving EQ 45–47, 56
sibilance 137, 167
sidechain compression 79–80, 120
sidechain EQ 80–81, 145
sine wave 91, 178
sinusoidal harmonic partial 189
sound design 144
spectrogram 189

SPL 3, 202
square wave 178
SSL 62, 68, 165, 204–205
standing wave 89
static mix 16, 34, 127
stem mastering 220–221
step sequencer 146, 185
stereo field 27, 29, 31, 33–34
stereo image 26, 29, 32, 96; and stereo enhancement 216; stereo imager 25–26, 28, 35
stereo width 135–136
sub-drop 144
subtractive/corrective EQ 49, 52–53, 57, 59, 163
synthesis 175; and additive 189; FM 187; granular 190; physical modelling 189–190; sample-based 187–188; spectral 189; subtractive 186–187; synthesiser 176; vector 188; wavetable 188; West Coast 190
synth sounds: bass 192; bell 191; brass 191; keys 191; lead 191; pad 191; pluck 191

tape hiss 144
tape loop 107
telephone EQ 54
threshold 65–66, 69
transient 66–67, 72, 76, 94; and transient shaper 79, 122
transition 140
triangle wave 178
trigger 183
true peak 211

unison 180
uplifter 140, 144

Variable Mu compression 69–70, 77–78
VCA compression 68, 76–77
velocity 145–146
vinyl 221–222
vinyl crackle 144
vocal throw 121, 143
voices 113–114, 180
VU meter 6–7, 14, 16, 20, 73

waveshape 178
white noise 185
wobble 109
wow 109

For Product Safety Concerns and Information please contact our EU representative GPSR@taylorandfrancis.com Taylor & Francis Verlag GmbH, Kaufingerstraße 24, 80331 München, Germany

Printed and bound by CPI Group (UK) Ltd, Croydon, CR0 4YY
08/06/2025
01897005-0004